Rethinking Innovation

Rethinking Innovation:
Global Perspectives

Editors

Ram Subramanian
Martin Rahe
Vishnuprasad Nagadevara
C. Jayachandran

Routledge
Taylor & Francis Group
LONDON AND NEW YORK

First published 2014 in India
by Routledge
711 Third Avenue, New York, NY 10017

Simultaneously published in the UK
by Routledge
2 Park Square, Milton Park, Abingdon, Oxon OX14 4RN

Routledge is an imprint of the Taylor & Francis Group, an informa business

First issued in paperback 2016

Typeset by
Bukprint India
B-180A, Guru Nanak Pura, Laxmi Nagar
Delhi 110 092

British Library Cataloguing-in-Publication Data
A catalogue record of this book is available from the British Library

ISBN: 978-0-415-74818-6 (hbk)
ISBN: 978-1-138-66023-6 (pbk)

Contents

List of Tables

List of Figures

List of Abbreviations

ACAP	Absorptive Capacity
APITD	Action Plan for Industrial Technology Development
CENIT	Consorcios Estratégicos Nacionales de Investigación Tecnológica
CIP	Competitive and Innovation Program
CRADAs	Cooperative Research and Development Agreements
CV	Curriculum Vitae
DARPA	Defense Advanced Research Projects Agency
ENTREPOR	Entrepreneurial Orientation
EOI	Export-oriented-industrialization
FAR	Facilitating Industrial Research Fund
FDI	Foreign Direct Investment
FP	Framework Program
GERD	Gross Expenditure on R&D
GT	Game Theory
HD-DVD	High Definition–Digital Video Disc
ICT	Information and Communications Technology
IE	Internet Explorer
IMP	Industrial Master Plan
IPO	Initial Public Offering
IPR	Intellectual Property Rights
IRPA	Intensification of Research in Priority Areas
ISI	Import-Substitution-Industrialization
IST	Information Society Technology
KPI	Key Performance Indicator
M&A	Merger and Acquisitions
MIUR	Ministry of Education, University and Research
MNCs	Multinational corporations
NAICS	North American Industry Classification System

NIEs	Newly Industrialized Economies
NPC	New Product Creativity
NPV	Net Present Value
NSF	National Science Foundation
OECD	Organisation for Economic Co-operation and Development
OEM	Original Equipment Manufacturing
OVALLMAT	Overall Maturity Level
PACAP	Potential Absorptive Capacity
RACAP	Realized Absorptive Capacity
RBV	Resource-based view
ROA	Real Options Analysis
ROI	Return on Investment
SBIRs	Small Business Innovation Researches
SME	Small and Medium Enterprises
TECHCAP	Technological Capabilities
WACC	Weighing Average of Cost of Capital
WTA	winner-take-all
VC	Venture Capital
VHS	Video Home System

Preface

This book is an outgrowth of the third Research Symposium (held in Barcelona, Spain, in June 2010) of the Society for Global Business and Economic Development (SGBED). The SGBED is a global academic organization that aims to foster cross-disciplinary and cross-cultural research from both developed and developing countries. In addition to its bi-annual conference, the SGBED conducts research symposia on a specific topic each year. The theme of the 2010 Barcelona conference was innovation and its many facets. Several scholars presented their papers and interacted with their peers who were interested in the topic of innovation. Of the many papers presented, the selected authors were asked to revise their papers in light of the symposium feedback. The revised papers were further reviewed by a panel of reviewers prior to inclusion in the current book.

The selected papers shed light on a wide variety of topics related to innovation. There are papers related to, among others, technology standards, team processes affecting innovation, and the interface between academia and industry. Collectively, these papers add to our understanding of what innovation is and the factors that impact it. The book's strength is the fact that innovation is approached from both the hard as well as the 'soft' (the people side of innovation) perspective. In addition, the book represents the current thinking on innovation by a global network of scholars, and so, it represents extant thinking on innovation from different cultural and economic contexts.

A multitude of people did yeoman work behind-the-scenes as well as in front centre to bring this book to fruition. Chief among them are Vishnuprasad Nagadevara, publisher relations, Martin Rahe, symposium organizer, and C. Jayachandran, President of the SGBED. In addition, the indefatigable Anthony Ndiege (of Montclair State University's Center for International Business) was joined this time around by fellow MBA student, Ram Siddharth Radhakrishnan, to bear the brunt of the day-to-day administrative details so necessary to produce the book. Our thanks to both of them.

May 2013

Ram Subramanian
Montclair, New Jersey, USA

Introduction

Michael Porter,[1] in his 1990 book, *The Competitive Advantage of Nations* stated clearly that prosperity is created and not inherited. Porter's concept of how to prosper is to create value for the society that generates income for the companies. However, the principal concern is how to sustain prosperity. The dynamics of global markets generate opportunities that may convert into competitive advantages on the one hand, but on the others they may erode rapidly, once competitors are able to catch up.

Economic theory has dealt with this issue for a long time. Harrod,[2] Domar[3] and Solow[4] analyzed the impact of knowledge and technology on economic growth, stating that a key element of sustained growth is progress in the improvement of human capital. Actually, this reflects in higher productivity and creates a competitive advantage. The principal objective of countries and companies focuses on the sustainability of competitive advantages, meaning that in dynamic environments the efforts to gain advantages against others is continuous. A good example is China, which shows how a poor country with cheap labour as its principal advantage, transforms itself to become a country with sophisticated manufacturing, higher income and an attractive domestic market. 'During the past several decades, Beijing, Shanghai, Shenzhen, and other cities have developed as centers in key information communication technology industries'.[5] China and other developing countries evolved as labour and consumer markets and in doing so, gained independence from the developed countries, and decoupled from their economic growth. This impressive development is, among other factors, the result of eager work on attracting technology and capital, benefitting from spill-over, and climbing up the production value chains by producing more sophisticated goods and services. Finally, the process requires willingness to change and the capability to innovate.

Today, innovation is the key driver of sustainable growth and economic prosperity. Having its origin in the Latin, *innovation* means *novelty* or *newness* — but new for whom? A single person, a company, an industry, a nation, or the whole world may consider something as new. This means that innovations are always context-related and the reference system for deciding whether something is new or not is the innovating subject. The implementation of changes in the corporate workflow is an innovation for the implementing company, but not necessarily for the competitor who is already operating for some time with these changes. Raising efficiency through the reduction of costs requires changes in the production process, and is a typical example of process innovation. A different type of innovation results from product diversification, and builds on the client as the reference system for evaluating the innovation success (product innovation). An example of an innovative product, which is objectively new because of its worldwide new technology application, is Apple's iPhone.

As Figure 1 shows, innovations take place within the environmental constraints resulting from legal, social, or technical norms, and have to consider state regulation and market pressures. As such, innovations are context-related and transform into process or product innovation, through the synthesis of the received input factors with the existing corporate configuration (technology, organization, staff).

To summarize, innovations result from environmental pressure (competition) and are evaluated by the markets (clients). Independent of the type of innovation, whether process-oriented or product-oriented, any innovation requires adjusting behaviour either in production or in consumer markets and is, therefore, a process of change.

Considering these circumstances, the manage-ment of innovation has turned out to become a key concern for private and public sector organizations. In its core, innovation management has a strategic and operational component. Strategy-wise, an organization has to plan on innovation. This includes a reflection on resources, and on the point of whether innovation capacity can be generated inside the organization with own resources, or through external support from innovation-experienced organizations.

Figure 1 System – Environment Model

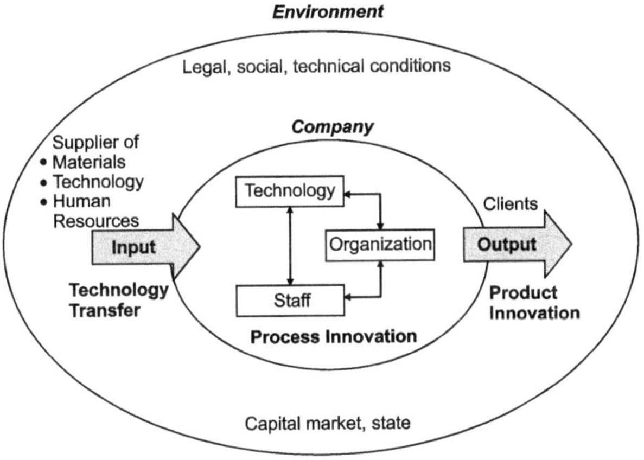

Source: Provided by the author.

Principal strategies are related to:

1. the implementation of a Research and Development (R&D) department,
2. the institutional support of intrapreneurship activities,
3. the creation of a joint venture, and
4. the search for an innovation alliance (innovation networks, triple helix).

The first two options are in-house strategies that require sufficient resources, especially financial, human, and time resources, while the other options are outside-oriented, trying to find appropriate partners that deliver missing capacities. Applying the make-or-buy distinction, the outside strategies can be regarded as buy decisions.

Often, innovation fails because of the company's incapacity to set up the right measures successfully. When talking about managing innovation, we should keep in mind that every innovation strategy has to be compatible with the organizational ability to manage the process. The operational task of innovation management is related to the following factors:

1. corporate configuration,

2. individual capabilities of the workforce,
3. timing, and
4. market issues.

Corporate configuration refers to those patterns that affect the organizational willingness to implement change, and includes aspects like organizational structure, culture, communication, relationships, etc. In this sense, the configuration is the medium that embeds concrete initiatives and policies on innovation. If the organization can rely on the capabilities of its workforce to support change, then tasks and responsibilities can be distributed. Here, the involvement of staff in innovation management leads to open minds, flexibility and trust. Configuration as well as capability are inside factors of innovation policies, while the other two aspects, timing and market, are external-oriented policies, and refer to the time-to-market span of innovative products and the perception of potential clients regarding the new product.

Looking at global trends, there is no doubt that the need for innovation is growing due to the increasing pressure from markets and environment. The challenge for governments, private and public organizations, and individuals will be how to manage the process in an economically, politically, and socially responsible way.

Martin Rahe
May 2013 **Barcelona, Spain**

Notes

1. Porter, M., *The Competitive Advantage of Nations*, New York: Macmillan, 1990.
2. Harrod, R. F., 'An Essay in Dynamic Theory', *Economic Journal*, vol. 49, 1939, pp. 14–33.
3. Domar, E. D., 'Capital Expansion, Rate of Growth, and Employment', *Econometrica,* vol. 14, 1946, pp. 137–47.
4. Solow, R. M. , 'A Contribution to the Theory of Economic Growth', *Quarterly Journal of Economics*, vol. 70, 1956, pp. 65–94.
5. Tan, Y. (2011), *Chinnovation: How Chinese Innovators are Changing the World*, New York: John Wiley & Sons, 2010.

1

Cluster as an Approach to Innovation Development of Entrepreneurships – A Case Study from Poland

Magdalena Bialic-Davendra and Drahomíra Pavelková

Innovations pose as a fundamental component of prosperous development of entrepreneurships. Clusters, through creating the possibility of cooperation among areas of science, technology, and economy, influence innovation development of enterprises collaborating within their structure. This article presents cluster as an environment-stimulating component in the field of research, and an innovation based on a case study from Poland. In addition, the new approach of cross-cluster cooperation has been introduced as an extended and diversified way, to attain high levels of innovativeness and competitiveness by collaborating entities.

Today's world wide processes of internationalization and globalization force entrepreneurships to search for new solutions in order to keep their position in the market and develop their activity. Innovativeness and integration are rising in importance as they consist of certain characteristics for survival in the world of competitiveness.

The concept of innovation plays a special role. The ability to introduce new innovation into the market is one of the most crucial conditions of competitiveness, though it is not an easy task. First and foremost, it requires the possession of a huge variety of resources such as natural resources, financial capital, human capital, etc. Innovation can be defined in a very broad way, starting from the one related to the product innovations regarding product or service, through process innovations introducing new technologies, to innovations in marketing or organization. However, the essence is to implement new solutions in practise.[1] The innovative solution can be defined as a result of its own research and development activity of an entrepreneurship, or as an effect of collaboration with other enterprises, institutions, etc. In addition, innovation can pose as an outcome of intangible knowledge such as patents, licenses, software, know-how, etc., or tangibles such as machines and devices. The main attribute is the fact that a new quality is being developed and introduced into practise.[2] According to Freeman and Sőlvell,[3] 'innovation is based on a process of incremental reduction of technical and economic uncertainty, where new technologies typically undergo a number of modifications and business models are adjusted accordingly. New particular knowledge and skills develop over time'.

Through introducing new products or services into the market, entrepreneurships try to respond to buyers' needs while simultaneously assuring for themselves a presence in the business. For the reasons of being more attractive to the customers and more effective in production, enterprises may also wish to implement new processes and/or activities in the company, change the technique or technology of production which has been in use, or the way the firm has thus far been operated. These activities pose different ways of implementing innovations within the company.

Enterprises have to be innovative in order to have competitive advantage above others. Innovations bring certain advantages for the company:

1. they influence the enterprise's performance, e.g., increase efficiency and productivity (affect productivity growth),
2. may result in increase of company's profitability,
3. increase chances of obtaining better position on the market and thereby help the firm to stay competitive.[4]

As competitiveness of entrepreneurships strongly depends on their level of innovativeness, companies feel the need to provide sophisticated methods and tools that enhance innovation technology development in the company.

In order to stay innovative, entrepreneurships, especially small and medium enterprises (SMEs) have to overcome certain difficulties. The two main obstacles are financial constraints and barriers connected with people's mentality. Financial barrier concerns very high cost of collaboration with research and development (R&D) institutions. These costs are hard to overcome for individual (small) companies, which often do not even possess a separate entity in charge of R&D in the business.[5] Secondly, the mental obstacle is connected with the fact that individual firms, very often, do not see the sense of spending large sums of money for activities which are incomprehensible for them, or actions of whose outcomes cannot be ascertained.[6] Extremely high expenditures on innovations and a limited number of prospective new solutions within the single units', force companies to take part in open innovations based on collaboration with external partners.[7]

The process of creating new innovations began to be realized within specified forms of cooperation, among different organizations from the areas of science, technology, and economy. Clusters play a special role among the innovation network linkages.

Cluster as an Approach to Innovation Development of Entrepreneurships

Literature adduces clusters in various ways. Nevertheless, the most well-known definition is the one presented by M. Porter,[8] stating that cluster is 'a geographically proximate group of interconnected companies and associated institutions in a particular field, linked by commonalities and complementarities'. All attributes are embedded within this definition. Close proximity is needed in order to enable occurrence of positive effects of mutual infiltration, and to benefit from the usage of the same resources. In addition, cluster gives an advantage of interconnection and interaction based on common aims, and the advantage of creation of networking linkages among entities from various areas.[9] Clusters link the phenomena of cooperation and competition with each other.

Clusters as a specific form of cooperation of industry and service sectors create the base for the development of creativity, innovativeness and competitiveness. They foster partnership activities among entities. Entities cooperating within a cluster create vertical and horizontal ties. A vertical tie represents the cooperation in the chain from supplier to final recipient of a particular initiative, while a horizontal tie involves cooperation among entities providing complementary products or services. Horizontal ties usually concern company-to-company relation or company-to-R&D sphere. These ties are the most important from the innovative point of view. Enterprises cooperating within a cluster may obtain certain benefits such as:

1. better and cheaper access to specialized inputs such as machinery, components, business services, qualified human capital resources, etc.,
2. great possibility of information, knowledge, know-how, and experience exchange,
3. receive faster, cheaper and more extensive information about the market, technological and technical developments, new marketing ideas, etc.,
4. opportunity to benefit from institutions and public goods enabling companies, e.g., to recruit already trained employees, to obtain (by lower costs) expert advice from the local institutions, to use specialized infrastructure, etc.

In addition, clusters stimulate competitive pressure even among indirectly competing or non-competing participants, which has a motivating influence on entities who thereby are more likely to co-locate R&D, cooperate on production, co-monitor and rapidly perceive customer needs, do common marketing activities, etc.[10]

Clusters stimulate and revitalize cooperation in the business environment. As highlighted earlier, entrepreneurships cooperating within such a network can gain multiple benefits derived from various activities.

Nevertheless, the most crucial area of cooperative activities seems to be oriented towards research and innovation development, as the way to stay competitive in the globalized market. For companies, it is easier and more economical to pool their efforts into common scientific research rather than doing them individually. The base for research and innovation development translates into the exchange and creation of new ideas. Thanks to the closeness of companies, there is a possibility of face-to-face contact. These, in turn, give firms a good opportunity to exchange experiences and knowledge, to learn from one another (co-learning), and learn by directly observing other firms (learn from their practice). Within a cluster, it is easier for companies to organize common workshops, traineeships, conferences, and other events, where people 'shuffle' their views and new ideas for further development are born, and where they can simply learn more (about investment programs, products certificates, and production processes; obtain knowledge about new technical and technological solutions, or current market trends, etc.).[11] Moreover, companies more readily collaborate on collective research and development projects as they have a trusting relationship with their partners. Common goals are achieved through mutual collaboration. Companies often share their equipment, and combine human resources (skills and qualifications of people). Through multiplying their resources, enterprises can experiment on lower costs, and attain more creative and original solutions. Furthermore, through mutual complementation, they reduce weak points or risks of a new innovation. Their synergy contributes to an increase in the effectiveness of the research, and proceeds in generating new results.

Collaboration of companies facilitates innovation progress not only in a cluster as a whole but also, especially in separate entities, which as isolated firms, would not have enough time nor money to lead research at high level, and to explore new innovative solutions.

Cooperation in the Area of Research and Innovations Based on an Examined Group of Polish Clusters

In Poland, the idea of clustering appeared in the 1990s and initially developed slowly. The interest in clusters increased gradually and since the year 2006, the importance of clusters was highlighted in governmental programs and strategies oriented towards innovation economy, innovativeness and competitiveness. Since no cluster policy has been founded so far, all activities directed towards clusters and cluster initiatives have been realized under innovation policy. The innovation policy in Poland supports linkages and networks of cooperation, joint ventures and investments of entrepreneurships, promotion of cooperation not only among enterprises but also within business surrounding institutions, exchange of information and experience, etc. It focuses on innovative activity resulting in creation and adaptation of new or improved solutions in products, services, and processes. The multidimensional and multithreaded character of innovation policy allows innovative development on different levels, starting from regional to national. In each voivodeship (province/ administrative division of Poland), respective Regional Innovation Strategies (RIS) were defined and implemented. The common aim of all strategies introduced, is to increase entrepreneurship within the region, increase competitiveness of scientific offers, utilize natural potentials, promote different types of co-operation — common planning of regional development (enterprise-science-economy networks), create innovative infrastructure, spread pro-innovative attitude, etc. All these activities create favourable background for clusters development. Clusters, in turn, pose the 'door' to progress of an individual company. They stimulate improvement, specialization, innovativeness, and competitiveness of their participants.

Within this research, a group of 32 Polish clusters were examined. The group characterizes a relatively young age; 29 out of 32 clusters (90.6 percent) were established during the last four years. The Tarrow Industrial Cluster (established in 1999) is the oldest cluster and the Vistula Renewable Energy Cluster (established in 2009) is the youngest one in the group. Figure 1.1 presents the clusters which were examined based on the location of the voivodeships.

Figure 1.1 Examined group of Polish Clusters According to the Voivodeships

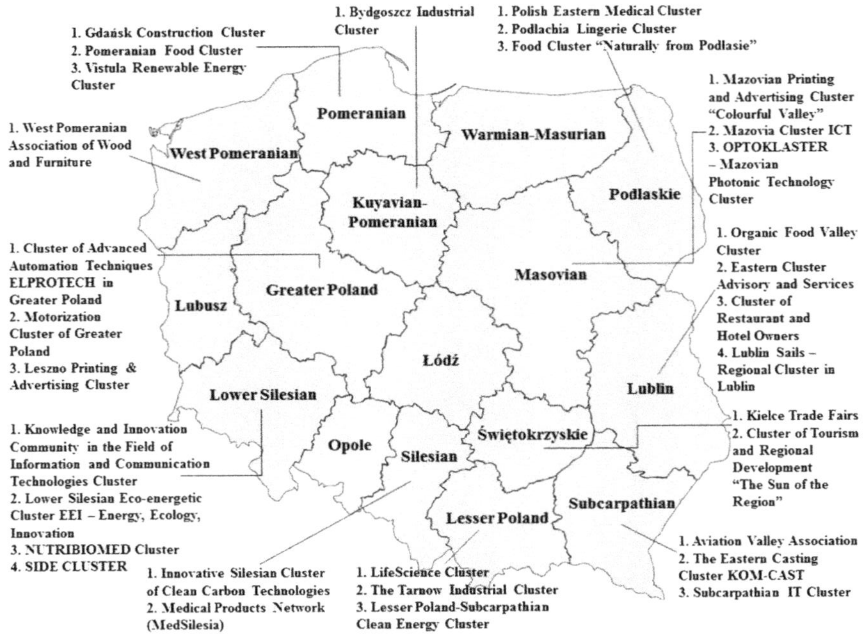

Source: Developed by the authors.
Note: The contours of the map of Poland adapted from http://bibioteka_zsg. manifo.com/mapy (accessed 6 August 2013).

These clusters lead their activity within a variety of industries starting from traditional branches — e.g., food production, wood industry, construction industry, clothes industry — to modern and high-technology industries — e.g., photonic industry, life science, IT, aviation, automation industries, etc. Disparity of industries in which the examined group of clusters operate, may render an impression that the particular clusters prioritize the main goal of their activity differently. However, this research found that the main goal of more than 60 percent of these clusters is to support entrepreneurships in a particular branch/industry, in order to help them maintain their innovativeness and competitiveness in the market, and to create thick bonds of cooperation among enterprises, science entities, governmental authorities (local and/or regional), and business surrounding institutions, in order to enable easier development of innovation processes and technology transfer.

As further goals, the following can be distinguished:

1. Realization of innovative projects within a particular branch;
2. Obtaining financial support from various sources, for example, EU funds, government subsidies, etc.;
3. Improvement of the competitiveness of SMEs in the region (spreading regional development based on particular production; comprehensive development of a region; promotion of region; job creation in the region);
4. Organization of workshops, conferences, symposiums, traineeships (increase of qualifications), and other events (exhibitions, trade fairs, festivals, etc.) where new information and knowledge can be exchanged;
5. Common promotion (common brand name creation);
6. Common production (an easier way of acquiring orders; costs reduction; obtaining new and/or better position on the conditions which dictate the market).

It is obvious that clusters create a natural environment for development. The principal developments being research activities, education and training, which in turn are inseparable parts of the innovation progress. Clusters give the opportunity to classify enterprises into workgroups, according to their specification of production or activity, that fosters more regular meetings within companies. This, in turn, creates a positive environment for discussion about common matters and problems, such as, common logistic activities focused on supplies, transport, warehousing, cooperation, and division in work, etc. Enterprises readily collaborate on the creation of collective cluster product, common unique offer for auctions (possible to present to wider scale of customers), and share orders. Thanks to cooperation, they are able to react faster to market changes, and undertake certain activities, leading to improvement in quality and competitiveness of products and/or services offered by the cluster. Together, they willingly participate in international events, which provides the opportunity to meet experts in that particular area of industry from all over the world. Furthermore, they are more open to the development of cooperation with foreign entities, that broadens companies' horizons in the fields of their activity.

Generally, a cluster goal is to lead cooperative companies to such organizational, technical, and technological changes which would enable them to produce on a high quality level, comparable with the best firms in the field.

The activities perceived by cluster managers as the most interesting for cluster members are prove to be complementary with key goals, posing the essence of cluster activity as a structure. Among the most attractive activities for cluster participants, managers identified the following (listed according to their popularity):

1. common promotion (common brand name creation),
2. spreading the network of collaboration (exchange of information, know-how, undertaking collective projects) and creation of cooperation bonds (enterprise-science-economy linkages),
3. organization of conferences, workshops and other events,
4. help in acquiring contacts (supplier-receiver), common orders, better conditions on the market,
5. lobbying in favour of financial support from authorities (subsidies, funds),

6. (a) common marketing, common offer on the market,
   (b) participation in international projects, trade fairs, economic missions,
7. (a) support of entrepreneurships in particular branch/ industry (facilitating innovative processes and technology transfer, integration of environment),
   (b) organization of networking (informal) meetings,
   (c) increase in human resource capital qualifications,
8. obtaining support for cluster activity,
9. creation of supportive institutions (technological parks, trade institutions),
10. (a) promotion of the region,
    (b) development of education at high school level,
    (c) support for spin-off creation.

For companies, initially, it is most crucial to build a cognisable brand name, in order to increase the consumers' awareness about certain products, services, etc. However, the importance of cooperation and its intensification is also high. Companies seem to understand the need of adhesion and cohesion in action. They realize how difficult it is for an individual company (especially SMEs) to survive in the market. Therefore, they decide to join their forces in common activities and ventures. Within a cluster, it is easier for enterprises to undertake collective research, work on common projects, and apply for support from authorities. It is also less difficult to maintain competitiveness due to the more open access to the latest research (e.g., technological foresight), analysis, and trends in the sector. All distinguished activities are complementary to one another, and directly or indirectly influence research and innovation development in enterprises.

Common ventures in the area of research and innovation are in-built in cluster activity. Clusters were asked which of the distinguished activities they currently offer; which they are planning to provide for cluster participants in the near future or upon request, and which they do not plan to do at all.

Figure 1.2 Joint Activities of a Cluster in the Area of Research and Innovations* (percentage)

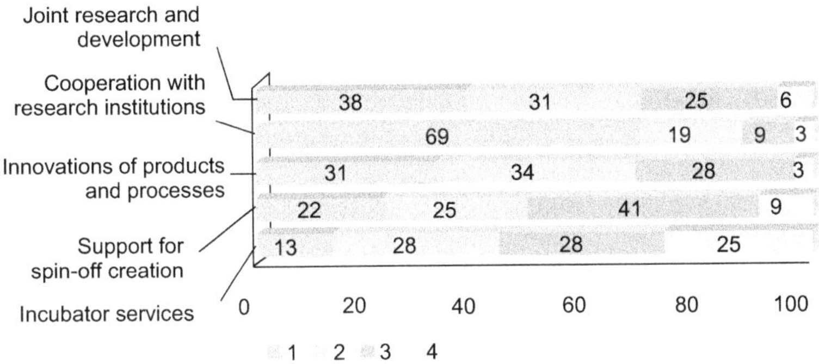

Source: Developed by the authors.

Notes: *Where: 1 – cluster currently offers activity; 2 – cluster currently does not offer activity but is planning to do so in a few years; 3 – cluster currently does not offer activity but is able to arrange it if needed; 4 – cluster neither offers nor plans activity.

As Figure 1.2 shows, the most interesting activity for the examined group was cooperation with research institutions (e.g., research and technological parks, universities, colleges, non-profit institutions, etc.), with 69 percent of clusters already providing the activity, and another 28 percent being ready to offer it. Thanks to this opportunity, cluster participants are able to pool their capital sources into common research projects making them more approachable (Table 1.1). Moreover, research institutions may also allow them to access the latest (used on the market) cutting-edge solutions, find out more information about the market itself (forecasting market behaviour), obtain expert opinion, or consult planned ventures. Very often, these amenities are not available to individual companies due to very high costs. Further in significance are joint R & D and joint innovations of products and processes, resulting in 38 percent and 31 percent respectively of already provided activities, with another 56 percent and 62 percent accordingly, being ready to offer, if required. Joint R&D allows companies not only to combine their knowledge and experience (since they often search for similar solutions in the area), but also to lower their expenditure spent on research (central management of research and development and division of costs). Moreover,

Table 1.1 Joint activities in the Area of Research and Innovations and their Benefits for Cluster Participants

Type of activity	*Benefits for cluster participants*
Joint research and development	Decrease in expenditure on research per entity Joining knowledge and experience Searching for common solutions in the same industry area Refining of exploit technologies Central management of research and development Finding new trends, areas, markets Composing and implementing innovative solutions available only for clusters and cluster initiatives (EU programmes, traineeship programs, etc.)
Cooperation with research institutions	Co-financing of research projects Direct access to research results realized by public institutions and universities e.g., market research Access to latest solutions Cooperation with experts in the area/consultancy Implementation of research results in practice
Innovations of products and processes	Cluster carries results of new technologies development and suggests the implementation of complex products and devices Faster and less risky implementation of innovations in the market Better conditions/position (of company) in the market Optimization of production processes
Support for spin-off creation	Enlargement of cluster Support of innovations in particular industry area, region Increase of flexibility in activity of a company (element of restructuring)
Incubator services	Development of entrepreneurship environment in the area

Source: Author's own conclusions[12] and from *Clusters and their Impact on Companies Performance*, Pavelkova, et al., p. 119, Prague: GRADA.

common research brings new findings about the market and new trends used within the industry which gives companies the opportunity to quickly respond to changes. In addition, some solutions provided (e.g., within special programmes) may only be available for companies collaborating within a cluster or cluster initiative. Regarding innovations of products and processes, cooperation within a cluster enables better recognition of market demand, and faster and less risky implementation of innovations in the market. Besides, enterprises applying modern solutions (e.g., advanced industrial technologies) and trends are more attractive to potential customers, and thereby they may reach an advantageous position, and obtain a chance to dictate conditions in the market. In the area of research and innovation activities, support for spin-offs and joint activities regarding incubator services are the least popular. Only 22 percent of clusters offer support in creating new entities derived from the parent company (even though, clusters do not initiate spin-off creations themselves, they offer them support in contacts creation, various discounts by virtue of being cluster members, etc.). Only 13 percent take action oriented towards incubator services (additionally, as many as one-quarter of examined clusters do not plan to offer this activity at all). Nevertheless, those activities may bring certain benefits as well, such as an expansion of the cluster or increase in the development of entrepreneurship environment in a particular area.

Clusters create a network of linkages among enterprises, science and economy. Adequate interaction among these three spheres is fundamental for the successful development of entities involved in any field, especially in the area of research and innovations.

Development of Innovativeness and the New Concept of Cross Cluster Collaboration — 'Clustering the Clusters'

Cluster, as a specific type of network, is a platform for cooperation of entrepreneurships, government authorities, educational, and business surrounding institutions, in particular, the geographical area. The need of proximity has provoked the creation of cooperation networks within the region or adjacent regions, so far. Nowadays, the new concept of 'clustering the clusters' appears. The meaning of the word 'cooperation' has been taken to a higher level. Cluster networks started to expand into partnerships, clubs, platforms of cooperation, alliances, bilateral agreements, trans-national cooperation, or groups of collaboration around the world. The innovativeness and competitiveness have been transferred from clusters to entire eco-systems. New opportunities for entrepreneurships cooperating within clusters have started to arise.

The need to improve the business and research environment has also been noticed within Polish clusters. Among the examined group, 69 percent of clusters (22 entities) declare their cooperation with at least one other cluster. The question arises, what are the reasons for this type of collaboration? Within the investigated group, 68 percent (15 of the clusters) cooperate in order to exchange knowledge, information, and experience in a particular branch/industry and models of collaboration. Another 27 percent collaborate on collective projects such as those related to EU or those concerning technology of production (foresight), etc. The other clusters are co-organizing educational events such as conferences, seminars and workshops, business-to-business meetings, or common events in character of promotion (advertisement of common products, promotion of region). Another group cooperates on collective ventures, activities (e.g., promotion), doing common research (preparation of innovative projects), or support the establishment of new emerging clusters. They usually show active cooperation in a few of the afore-mentioned areas simultaneously.

Whole cluster structures as well as single enterprises feel 'the weight of competitiveness', which is why they resort to new solutions such as cluster-to-cluster cooperation. Through they focus on mutual exchange, policy-learning, and collaboration which in turn contributes to the progress of each member. Mutual motivation and stimulation help to reach the level of world-class performance. Single entities realize that in order to 'be in the running' in a global market, they have to possess concrete tools to influence the transfer of knowledge and innovative activities. Since those tools are often unattainable for them as individuals, they see cooperation and partnership within the cluster structure as a way of extension of business affiliations, bringing competitive advantages and further development.

The concept of cross-cluster cooperation, the reason behind requirements and/or restrictions for this type of collaboration, as well as its influence on cluster structures and the performance of single enterprises are a subject for further investigation.

Conclusion

Being innovative is a hallmark of economic dominance and successful performance of an entrepreneurship. As individuals, companies have very limited resources and often cannot afford to be innovative. Therefore, they are motivated to participate in open innovations, based on cooperation with external partners. As a good approach to research and innovation development, the collaboration within cluster network appears. Cluster as a structure provides a series of possibilities which influence the progress of its participants. Thus, innovation development could be supported mainly by the following activities:

1. networking,
2. mutual communication among companies,
3. access to information,
4. access to innovative technology,
5. joint research/cooperation with research institutions,
6. education of human resources, and
7. cooperation with educational institutions.

Enterprises know that innovations are based not only on physical resources (access to laboratories, expensive equipment, huge amount of capital supporting research) but first and foremost on human capital and people's ideas (skills and qualifications plays a crucial role in innovative processes). Innovation processes within a cluster are based on continuous interaction across organizations. Clusters, as a specific type of network, assure the opportunity for information exchange, funds conjunction, expertise exchange, and ordered and diversified cooperation among entities, providing fundamentals for innovative development. Companies, through multiplying their resources, can obtain more creative and original solutions. The possibility for cross-cluster cooperation extends the horizon of the entities even more, and it becomes a new perspective for building up knowledge and experience for companies, leading to innovative solutions on a wider scale.

The experience of clusters confirms the importance of dynamic development in the area of research and innovations for the healthy progress of clusters themselves and all its members.

Notes

*The authors are thankful to the International Visegrad Fund No. 50910431, the Internal Grant Agency of FAME TBU No. D/09/5, and the Internal Grant Agency of FAME TBU No.DP 9/10 for financial support to carry out this research.

1. Daszkiewicz, M., *Jednostkibadawczo-rozwojowejakoźródłoinnowacyjnościwgospodarceipomocdlamałychiśrednichprzedsiębiorstw* (Research and Development units as a Source of Innovation in the Economy and support for Small and Medium-sized Enterprises), Warsaw: PARP, 2008.
2. Daszkiewicz, M., *Jednostkibadawczo-rozwojowejakoźródłoinnowacyjnościwgospodarceipomocdlamałychiśrednichprzedsiębiorstw* (Research and Development units as a Source of Innovation in the Economy and support for Small and Medium-sized Enterprises), Warsaw: PARP, 2008.
3. Freeman, C., *The Economics of Industrial Innovation*, second edition, London, Frances Printer Publishers, 1982; Sőlvell, Ő., *Clusters — Balancing Evolutionary and Constructive Forces*, Second edition, Ődeshőg: DanagårdsGrafiska (ŐdeshőgDanagårds Graphic), 2009.
4. Szymura-Tyc, M., 'Innowacja wartości w międzynarodowej sieci przedsiębiorstw' (Innovation of value in the international network of companies), in H. Brdulak and E. Duliniec & T. Gołębiowski (eds), *Wspólna Europa. Partnerstwo przedsiębiorstw jako czynnik ograniczania ryzyka działalności gospodarczej* (Common Europe. Partnership of firms as a factor in mitigation of risk in economic activity), Warsaw: SGH, 2009, 341–51; Wojnicka, E., 'Interakcje w procesieinnowacyjnymjakoczynnikkonkurencyjnościprzedsiębiorstw' (Interaction in the innovation process as a factor of

enterprises' competitiveness) in M. Górzyński (ed.), *System wsp ieraniagronprzedsiębiorczości – publikacjapodsumowująca* (The system of support for business clusters - summarizing publication), Rzeszów: WSIiZ, 2006, 9–32.

5. Szultka, S., *Klastry Innowacyjne wyzwanie dla Polski*, (Clusters Innovative Challenge for Poland) Gdańsk: IBnGR, 2004.
6. Daszkiewicz, M., *Jednostkibadawczo-rozwojowejakoźródłoinnow acyjności w gospodarceipomocdlamałychiśrednichprzedsiębiorst w*(Research and Development units as a Source of Innovation in the Economy and support for Small and Medium-sized Enterprises), Warsaw: PARP, 2008.
7. Pełka, W., 'Nowe formy współpracy w zakresiedziałalnościinnowacyjnej firm' (New Forms of Cooperation in the Field of Firms Innovation Activity), in H. Brdulak, E. Duliniec and T. Gołębiowski (eds), *Wspólna Europa. Partnerstwo przedsiębiorstw jako czynnik ograniczania ryzyka działalności gospodarczej* Common Europe. Partnership of firms as a factor in mitigation of risk in economic activity, Warsaw: SGH, 2009, 277–86.
8. Porter, M., *On Competition*, USA: *Harvard Business Review Book*, 2008.
9. Gorynia, M., and B. Jankowska, *Klastry a międzynarodowako nkurencyjnośćiinternacjonalizacjaprzedsiębiorstwa* (Clusters and International Competitiveness and Internationalization of an Enterprise), Warsaw: Difin, 2008.
10. Porter, M., *On Competition*, USA: *Harvard Business Review Book*, 2008.
11. Łęcznar, M. 'Znaczenieklastrów w transferzewiedzyiinnowacji – wnioskidlaPodkarpacia' (The Importance of Clusters in Transfer of Knowledge and Innovation – Conclusions for the SubcarpathianVoivodeship), http://www.rsi.podkarpackie.pl/praktyki/publikacje/zalesie/artykuly/125%20Lecznar.pdf (accessed 22 January 2010).
12. Pavelková, D., L. Friedel, E. Jirčíková, A. Knápková, K. Skokan, and P. Škodáková, *Klastry a jejichvlivnavýkonnostfirem* (Clusters and their Impact on Companies' Performance), Prague: GRADA, 2009.

2

New Product Creativity and Competitive Advantage: Relevancy to the Malaysian Economy

Pelin Bicen and William H.A. Johnson

Technological advancement and the creation of new and creative products are key driving forces of economic growth. Many studies have measured the relationship of creative and innovative products with the firms' superior financial performances.[1] The results have mostly confirmed that creative products account for the lion's share of the firms' superior financial performance and growth. Interestingly, most of these studies have been set within the context of advanced industrial economies operating at the frontiers of technology, in particular the United States of America (USA). Despite the centrality of creative products to a firm's competitive advantage and therefore its survival, the innovation management literature has given relatively little attention to the understanding of mechanisms of New Product Creativity (NPC) in Newly Industrialized Economies (NIEs). This is despite the fact that NIEs have grown so rapidly over the past forty years.[2] While firms in NIEs are by no means neglected, the weight of the discussion has so far focused on understanding NPC processes in the firms in Western culture.

In this article we discuss the essentiality of NPC and its impact on the firms' competitive advantage in NIEs, focusing on Malaysia. That is, we ask whether the NPC concept is relevant to the Malaysian context. While the macro and micro economic variables in Malaysia are rather different from what exists in advanced industrial economies, we argue that because competition takes place on a global scale, the processes of developing unique and meaningful products and their importance to the firms' competitive advantages have certain similarities. Therefore, the purpose of this paper is twofold. First, we discuss the relationship between new product creativity and competitive advantage. Second, and more importantly, we discuss the idea that their relationship may, in fact, be more general. For example, because firm-level capabilities of accumulation and assimilation leads to technological development of industries and economies,[3] *we explore the role and importance of new product creativity for competitive advantage at both the firm and national levels of analysis. That is, we argue for the importance of NPC specifically to Malaysian firms as well as for Malaysia as a nation, in general. The analysis of national policies of technological progress suggests the move towards developing NPC at the firm level, and that this ultimately affects the national competitiveness of Malaysia. Finally, we provide implications for this perspective in the concluding section of the chapter.*

The marketing and management literatures have paid significant attention to the concept of NPC and its contribution to the firms' competitive advantages.[4] In this stream of research, NPC has been defined in different ways. In this study, we adopt the Im and Workman[5] definition of NPC for two reasons. First, drawing from research in the fields of psychology, marketing, and organizational behaviour, they provided a detailed background of the concept of NPC. Their research is the most up-to-date and extensive study regarding this concept. They defined NPC as 'the degree to which the new products are perceived to represent unique differences from competitors' products in ways meaningful to target markets'. Second, and consistent with Amabile[6], they use the output perspective of creativity which includes two distinct sub-dimensions of the construct, differentiated in terms of novelty and meaningfulness. In the current study, we examine the importance of NPC for the Malaysian economy and its economic growth. Following Im and Workman[7] and Amabile[8], in our examination, we treat NPC as an towards consumer-based, high-tech products as a response to global-scale competition. In this shift, Malaysia's technology policy has played a significant role. Accordingly, three periods in the evolution of technology policy are evident:[21] (1) Independence to the early 1980s, (2) from the mid-1980s to early 1990s, and (3) from the mid-1990s to the present. Having a colonial history, Malaysia's post-Independence economic efforts focused mostly on agricultural development. Support for Malay dominant agriculture and local universities for conducting research in agricultural sciences were given the highest priority in the country's industrial policy. Technological development was of relatively low priority in Malaysia's agenda for economic progress. Instead, economic policy mainly focused on an import-substitution-industrialization[22] (ISI) model, which eventually led to capital-intensive production and low employment generation.[23] Criticisms about ISI policies convinced the government to shift towards executing an export-oriented-industrialization[24] (EOI) model, which attracted foreign direct investment (FDI) from many multinational corporations (MNCs). Early technology policies were mainly passive and restrictive, focusing mostly on regulating technology acquisition rather than a technology assimilation process of MNCs' product development initiatives. Therefore, the spillover effect of technological development to the local business at this point was very limited.

It was not until administrative changes in the Malaysian government took place in the early 1980s that it was realized, that there should be more emphasis on development of creative new products to compete effectively in the international arena. In mid-1980s, the government's first concrete plan, the Industrial Master Plan (IMP), attempted to actively involve local businesses in developing technologically advanced products. Executing this plan led to the government learning that building up technological competence was hardly the objective of many local companies. Instead, it was mainly the foreign MNCs that were investing in high-tech industries (e.g., electronics and semiconductors), and these industries depended almost exclusively on technology acquisition.[25] This new awareness of the Malaysian government accelerated state initiatives, to take significant policy measures towards building indigenous technological innovation capabilities. One such initiative was the Intensification of Research in Priority Areas (IRPA) programme enacted in 1986. The IRPA programme aimed to streamline public Research and Development (R&D) funding and investment, towards developing selective industries as well as encourage greater interaction between public institutions and private-sector industry representatives. The impetus of IRPA encouraged the creation of the 1990 Action Plan for Industrial Technology Development (APITD) and, eventually, the reform in the National Science Council. Thus, the Action Plan and the reformed National Science Council motivated policy makers to build new policy-making bodies having less centralized and formalized structures.[26]

These less bureaucratic institutions aimed to bond stronger links between government officials and selective private-sector representatives, to accelerate the technological progress required for developing creative new products in certain areas. These areas included the following industries: automated manufacturing technology, advanced materials, biotechnology, electronics and information technology, energy, environmental and aerospace technology. The early 1990s' reforms reflected significant growth in Science & Technology indicators (Table 2.1). Under the Sixth Malaysian Plan (1991–1995), high-tech exports ratio to manufactured exports rose by 4.4 percent, residential patent applications rose by 21.4 percent, and residential trademark applications rose by 56.4 percent of the total.

Table 2.1 Science & Technology (S&T) Indicators between 1991 and 1995

	1991	*1992*	*1993*	*1994*	*1995*
High-tech exports (percent manufactured exports)	17.9	17.3	17.9	18.4	18.7
Patent applications	551,201	611,774	629,015	617,881	669,065
Trademark applications	848,153	919,917	972,422	1,077,827	1,157,559

Source: World Bank Report (2005), http://info.worldbank.org/etools/docs/library/205682/7%20Dato%20Kaziah%20%20Session%207%20Plenary%20Incentives.pdf (accessed 5 August 2013).

Interestingly, the government's effort to build a demand-driven technology infrastructure by focusing more on private sector needs, unexpectedly turned into more state-led technology policy initiatives and practices. Mahathir bin Mohamad, who had served as the Prime Minister of Malaysia from 1981 to 2003, been in power over the entire course of policy changes aimed at increasing the technological capabilities of Malaysia. Early on, the Mahathir administration's main strategy was to motivate the private sector (i.e., by creating a demand-driven infrastructure) to build international technology capabilities, and develop creative products by providing Malayan businesses with incentives, and building technology institutions to assist them with their industrial development goals. However, the administration's concentrated effort to reform the technology policies turned into a more centralized supply-driven infrastructure.[27] Though this was not what the Mahathir administration was aiming for, it did make a contribution to the progress of public R&D infrastructure. In another attempt to strengthen the demand-driven technology infrastructure, the state encouraged MNCs to invest in export-oriented high-tech industries. Encouragement came in different ways such as tax breaks, allowing 100 percent equity ownership and exemptions from various business laws.[28] Though the administration's main expectation from MNCs was to invest in public technology institutions and, therefore, develop the indigenous technology capabilities of the nation, it did not turn out that way. However, opportunity came knocking in the mid-1990s. Due to globalization and free trade practices, MNCs and their subsidiaries became more competitive and dynamic, especially in high-tech manufacturing industries. Eventually, this encouraged them to forge closer linkages with local vendors of components.[29] Briefly, challenges created by global competition turned into a challenge for local businesses to be more integrated, and develop sub-parts and components required for developing competitive high-tech products. This opportunity changed the technology policies to emphasize building competitive industrial clusters,[30] launching vendor development programs, and training local vendors in relationship-building with MNCs.

An emphasis on industrial clusters as a way to technological advancement and developing creative new products was fully articulated in the Seventh Malaysian Plan (1996–2000), Second Industrial Master Plan (1996–2005), and Vision 2020 as a central principle in guiding technology policy.[31] In sum, the dominant theme in Malaysia's technology policies has been the synergistic collaboration between a model that lays emphasis on the private-sector institutions and liberalized economy, and a model that underlies the state's strategic intervention as a facilitator in MNC-led strategy. It is evident that in the process of Malaysia's industrial technology development, the state's adaptive and constructive policies have played a significant role in the upgradation and development of internationally competitive creative new products for Malayan firms.

Based on the aforementioned discussion on technology policy evolution in Malaysia, we argue that assimilation theories provide a foundation for Malaysia's transition to technology-based industrialization. They also provide reasoning for the role and importance of creative new products in developing Malaysia's competitive advantage. According to neo-classical economists, the success of technologically advanced nations relies on the high investment rates in human and physical capital.[32] They argue that high investment rates in tangible resources enable economies to move along a production function.[33] Nelson and Pack[34] call these arguments as 'accumulation theories'. These theories provide an explanation for the capital-intensive modes of high-tech production, which is considered an automatic consequence of such investments. In contrast, Austrian economists argue that entrepreneurship, innovation, and organizational learning, all supported and encouraged by technology policies, are central to technological advancement and the economic growth of a nation. They further argue that although human and physical capital is a necessary ingredient for technological progress, they are far from being sufficient to meet the long-term objectives for progress and provide a competitive edge to economies.[35] Rather, the creation of intangible resources and capabilities towards proactive innovation are essential to creating competitive advantages.[36] Nelson and Pack[37] denote these views as 'assimilation theories'.

Based on the systematic evaluation of Malaysia's technology policies, we argue that Malaysia changed its focus from technology accumulation to technology assimilation in the mid-1990s. Before the mid-1990s, virtually the entire increase in Malaysian exports was represented by products for which Malaysian firms had no insight into how to develop on their own. It was mainly the foreign MNCs who were leading the rapid growth in output and export. Briefly, Malaysian technology policies were mostly externally driven, and a technology-less form of industrialization. As a result, learning and spillover effects from the FDI of foreign MNCs were minimal. However, when competition became intensely global in the mid-1990s, Malaysia changed its approach towards technology-assimilation practices. The Malaysian government's main intention was to create a business environment where risk-taking and gaining entrepreneurial skills were appreciated and rewarded. In that sense, the government's efforts to stabilize the macroeconomic environment; to provide strong incentives to firms developing creative products; and, to form public institutions to create demand-side technology reform — all accelerated technology assimilation.[38] Figure 2.2 illustrates how Malaysia's high-technology export increased during the technology-policy changes that promoted technology assimilation and indigenous growth during this period (1996–2000).

Figure 2.2 High-technology exports (of manufactured exports) between 1996 and 2000

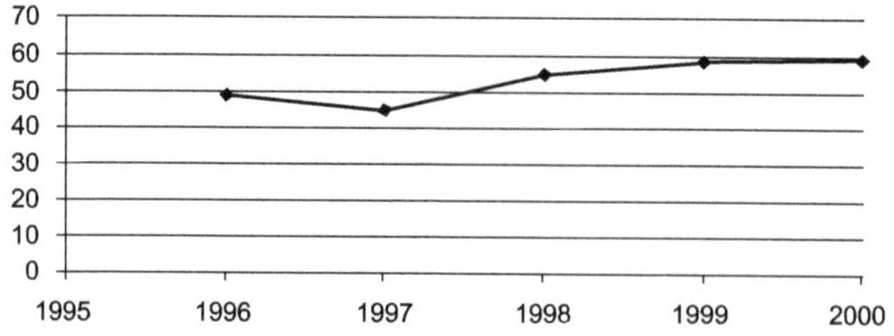

Source: World Bank Report

This statist political economy approach of Malaysia has paid off by deeply shifting Malaysia's focus on innovation in general, and leading Malaysian firms to develop unique and meaningful products to compete in the global arena in particular. This paradigm shift in Malaysian technology policies has reflected positively on performance for the country's exports. Table 2.2 summarizes Malaysian export trends between 1960–2000. Commodities (e.g., rubber, agricultural products) originally played an essential role in the Malaysian economy in the post-Independence era, but this trend changed over time, so that commodities accounted for only 20 percent of the overall exports as Malaysia moved into the 21st century.

Table 2.2 Export Trend in Malaysia 1960–2000 (percentage)

Year	*Commodities*	*Manufacturers*
1960	94	6
1970	92	8
1979	82	18
1991	39	61
2000	20	80

Source: Tradechakra.com, Tradechakra,http://www/tradechakra.com/economy/malaysia/malaysia-exports-173.php.(accessed 5 August 2013).

On an even more positive note, according to the recent 2010 report of Malaysia's trade statistics, the country is now working on next-generation products (e.g., semi-conductors, electrical and electronic products) and they account for 40.8 percent of the overall exports (Table 2.3).

Table 2.3 Major Export Products in Malaysia in 2010

Major Export Products	*Percentage*
Electrical and electronic products	40.8
Palm oil	7.6
Liquidated natural gas	6.5
Chemicals and chemical products	6.1
Crude petroleum	4.7
Refined petroleum products	3.6
Machinery, appliances, and parts	3.1
Manufactures of metal	3.0
Optical and scientific equipment	2.7
Rubber products	2.4

Source: Malaysia External Trade Statistics, 2010.

In sum, current adaptive and supportive technology policies in Malaysia, make it possible for firms to assess their technological strengths and weaknesses, search for, and evaluate external technology support, and integrate their firm-based resources to complement and enrich their technology capabilities. All of these synergistically combine to develop unique and meaningful products that can effectively and efficiently compete in global markets. The important point to note is that the development of a national capability toward innovation and technology development capabilities, based within industry, requires both state-level policies that support business and the entrepreneurship of creative and unique businesses themselves. As such, creating unique and meaningful products depends on investments and a market environment that facilitates efficient allocation, though the investments required for developing creative products also need to be induced by entrepreneurial firms.[39]

Conclusion

Contrary to older models of innovation, where innovation progresses in a linear and sequential fashion, from concept development to market commercialization, new models of the NPD process are more interactive, partly endogenous, and fundamentally uncertain.[40] New models integrate both market-pull and science-push approaches. Further, contrary to old innovation models where resources are seen as factor endowments, new innovation models, describe resources necessary for developing creative new products as more significantly heterogeneous across firms, intangible, imperfectly tradable, and imperfectly mobile.[41] When the fuzzy nature of the new innovation models are taken into account, it is harder to make sense of accumulation theories of innovation, where the central impetus for development lies in investments in physical and human capital, and where innovation is considered to be a by-product of these investments.

In this article, we have argued regarding two issues: firstly, that new product creativity is an antecedent for firms' competitive advantages, and that the accumulation of firm-level competitive advantages is essential in corporate growth strategies, which ultimately propels the competitive edge of nations; and, (2) drawing on assimilation theories of innovation, that technology policies play a significant role in encouraging and supporting local firms to develop unique and meaningful products. In that sense, state efforts to induce firms to export and compete effectively in world markets have facilitated and stimulated local firms to gain entrepreneurial skills, accelerate their organizational learning, and leverage their risk-taking abilities in R&D efforts. In this article, we specifically examined the dynamics inherent in the development of innovation and technological strategies in Malaysia. The reason for our choice of this country is twofold. Firstly, Malaysia is one of the strongest nations among the world's newly industrialized economies. Indeed, Malaysia's international competitiveness in high-tech products has augmented with the adoption of export-oriented industrialization (EOI) policies. As Lall[42] argues, the Malaysian government's effective policy interventions, enforce local companies to strengthen their R&D skills, increase their interaction with MNCs to assimilate state-of-the-art technologies, increase their risk-taking entrepreneurial skills, and, eventually, their innovativeness. Secondly, Malaysia has come far very quickly — from agricultural products which need low-level tech-skills to semi-conductors and electronics which need sophisticated high-tech capabilities, from a technologically backward economy to an affluent one and from a country of farmers to one of innovators and knowledge managers.

If NPC and the gale of creative destruction are the new realities of the 21st century's business environment, based on the analysis here, we can argue that Malaysia has so far met that challenge, especially at a stage where it is making a pivotal transition to a high-value added and industrially-oriented economy.

Our analysis implies that new-product creativity is a universally essential concept in the economic growth of nations. However, interestingly, discussion about new-product creativity, its developed measures, and conducted studies, that examine the antecedents and consequences of this construct have so far focused primarily only on Western companies. Therefore, we encourage researchers to further examine the NPC concept and its implications in NIEs, validate its measures, and conduct empirical studies that examine the antecedents and consequences of NPC, and evaluate whether the results are universal. We hope that our discussion here presents a modest beginning to such a research agenda and motivates further research to examine this interesting phenomenon.

Notes

1. Moorman, C., 'Organizational Market Information Processes: Cultural Antecedents and New Product Outcomes'. *Journal of Marketing Research,* vol. 32 (August), 1995, 318–35; Moorman, C. and Miner A.S. 'The Role of Organizational Memory in New Product Performance and Creativity', *Journal of Marketing Research,* vol. 34 (February), 1997, pp. 91–106; Rindfleisch, A. and Moorman, C., 'The Acquisition and Utilization of Information in New Product Alliances: Strength of Ties Perspective',*Journal of Marketing,* 65 (April), 2001, 1–18; Im, S., & J. P. Workman, 'Market Orientation, Creativity, and New Product Performance in High-Technology Firms', *Journal of Marketing,* 68 (April), 2004, 114–32; Sethi, R., D.S. Smith and C.W. Park, 'Cross-Functional Product Development Teams, Creativity, and the Innovativeness of New Consumer Products', *Journal of Marketing Research,* 38 (February), 2001, 73–85.
2. Kim, L., *Imitation to Innovation,*Boston: Harvard Business School Press, 1997.
3. Hobday, M., 'East Asian Latecomer Firms: Learning the Technology of Electronics', *World Development,* 23 (7) 1995, 1171–93.
4. Rindfleisch and Moorman. 'The Acquisition and Utilization of Information in New Product Alliances: Strength of Ties Perspective'; Im, and Workman, 'Market Orientation, Creativity, and New Product Performance in High-Technology Firms'; Sethi, R., D.S. Smith and C.W. Park, 'Cross-Functional Product Development Teams, Creativity, and the Innovativeness of New Consumer Products'; Brockman, B. K. and R. M. Morgan, 'The Role of Existing Knowledge in New Product Innovativeness and Performance', *Decision Sciences,* 34(2), 2003,

385–419; Brockman, B. K. and R. M. Morgan, 'The Moderating Effect of Organizational Cohesiveness in Knowledge Use and New Product Development', *Journal of the Academy of Marketing Science*, 34(3), 2006, 295–308; Cooper, R. G. (1979). 'The Dimensions of Industrial New Product Success and Failure'. *Journal of Marketing,* 43 (Summer), 1979, 93–103; Porter, M. E., 'What is Strategy?', *Harvard Business Review,* 74 (November–December), 1996, 61–78.

5. Im and Workman, 'Market Orientation, Creativity, and New Product Performance in High-Technology Firms'.
6. Amabile, T. M., 'The Social Psychology of Creativity: A Component Conceptualization', *Journal of Personality and Social Psycholozy*, 45(2) 1983, 357–376.
7. Im and Workman, 'Market Orientation, Creativity, and New Product Performance in High-Technology Firms'.
8. Amabile, 'The Social Psychology of Creativity: A Component Conceptualization'.
9. Sethi, R., D.S. Smith and C.W. Park, 'Cross-Functional Product Development Teams, Creativity, and the Innovativeness of New Consumer Products'. Prahalad, C. K. and G. Hamel, 'The Core Competence of the Corporation', *Harvard Business Review,* (March/April), 1990, 79–91.
10. Porter, 'What is Strategy?'.
11. Hunt, S. D., *A General Theory of Competition: Resources, Competences, Productivity, Economic Growth,*California: Sage Publications, 2000.
12. Day, G. S., 'The Capabilities of Market Driven Organizations', *Journal of Marketing,* 58(4) 1994, 37–52.
13. Day, G. S., 'The Capabilities of Market Driven Organizations', *Journal of Marketing,* 58(4) 1994, 37–52. Capabilities are defined as 'complex bundles of skills and accumulated knowledge, exercised through organizational processes that enable firms to coordinate activities and make use of their assets'.
14. The details can be found at http://www.marketwatch.com/story/microsofts-zune-continues-to-struggle (accessed 29 July 2009).
15. Barney, J., 'Firm Resources and Competitive Advantage', *Journal of Management,* 17(1), 1991, 99–120.
16. Hunt, *A General Theory of Competition: Resources, Competences, Productivity, Economic Growth.*

17. Hunt, S. D. and R. M. Morgan, 'Resource-Advantage Theory: A Snake Swallowing Its Tail or a General Theory of Competition?', *Journal of Marketing,* 61 (October), 1997, 74–82.
18. The details can be found at http://getclicky.com/marketshare/us/droid/(accessed 29 August 2013.)
19. According to the World Bank 2007 Trade Report, Malaysia is the third largest economy in the South East Asia and the 29th largest economy in the world by purchasing power parity with GDP.
20. Lall, S., 'Technology Policy and Competiti-veness in Malaysia', (eds) G. Felker and Jomo K. S. *Technology, Competitiveness and the State: Malaysia's Industrial Technology Policies,*London: Routledge, 1999.
21. Felker, G. and Jomo K. S., 'Introduction'. (eds) G. Felker and J. K. S. *Technology, Competitiveness and the State: Malaysia's Industrial Technology Policies,* London: Routledge, 1999.
22. In the developmental economics literature, ISI is described as a government's effort to reduce the country's foreign dependency through domestic production of industrialized products. In that sense, local companies are protected from international competition with trade barriers.
23. Felker, and Jomo 'Introduction', 1999; K. S., Jomo and C. Edwards, 'Malaysian Industrialization in Historical Perspective' in K. S. Jomo, *Industrializing Malaysia: Policy, Performance, Prospects*, London: Routledge, 1993.
24. In developmental economics, EOI policy is described as an economic policy that aims to advance industrialization process through export of goods for which the nation has a comparative advantage. In that sense, the main government policies were mostly made up of fiscal incentives tied to export performance, free-trade zones created to attract FDI, lowering trade barriers, and tariff rebate systems.
25. MIDA/UNIDO, 'Medium and Long Term Industrial Master Plan: Malaysia 1986–1995', 3(6), 1986, R&D and Technology Policies, Kuala Lumpur: Malaysian Industrial Development Authority;
26. Felker, G. and Jomo K. S. ,'Introduction' in G. Felker and J. K. S. (eds),*Technology, Competitiveness and the State: Malaysia's Industrial Technology Policies*, London: Routledge, 1999.
27. Felker, G. and Jomo K. S. ,'Introduction' in G. Felker and J. K. S. (eds),*Technology, Competitiveness and the State: Malaysia's Industrial Technology Policies*,London: Routledge, 1999.

28. Felker, G. and Jomo K. S. ,'Introduction' in G. Felker and J. K. S. (eds), *Technology, Competitiveness and the State: Malaysia's Industrial Technology Policies*, London: Routledge, 1999.
29. Rasiah, R., 'Flexible Production Systems and Local Machine-tool Subcontracting: Electronics Components Transnationals in Malaysia', *Cambridge Journal of Economics,* 18, 1994, 279–98.
30. Porter, M.E., 'Competitive Advantage', New York: The Free Press, 1990.
31. Lall, 'Technology Policy and Competiti veness in Malaysia', Felker and Jomo 'Introduction'.
32. Kim, J. I. and L. J. Lau, 'The Sources of Economic Growth in the East Asian Newly Industrialized Countries', *Journal of Japanese and International Economics,* 8 (September), 1994, 235–71.
33. Krugman, P., 'The Myth of Asia's Miracle', *Foreign Affairs,* 73(6) 1994, 62–78.
34. Nelson, R. R. and H. Pack, 'The Asian Miracle and Modern Growth Theory', *The Economic Journal,* 109, 1999, 416–36.
35. Kim, 'Imitation to Innovation',
36. Johnson, W.H.A., 'Transitions in innovation: musings on the propensity and factors towards proactive innovation in China', Jain, S.C. (ed.), *Emerging Economies and the Transformation of International Business: Brazil. Russia, India and China,* Cheltenham: Edward Elgar, 2006.
37. Nelson and Pack, 'The Asian Miracle and Modern Growth Theory'.
38. World Bank, 'The East Asian Miracle: Economic Growth and Public Policy', New York: Oxford University Press, 1993.
39. Nelson and Pack, 'The Asian Miracle and Modern Growth Theory'.
40. Felker and Jomo 'Introduction'.
41. Hunt, *A General Theory of Competition: Resources, Competences, Productivity, Economic Growth.*
42. Lall, 'Technology Policy and Competiti-veness in Malaysia'.
43. Lall, S., 'Technology Policy and Competiti-veness in Malaysia', (eds.) G. Felker and Jomo K. S. *Technology, Competitiveness and the State: Malaysia'a Industrial Technology Policies.* London: Routledge, 1999.

3

Option-Games Analysis on the Strategic Partnership of Biotech Start-ups

Takao Fujiwara

For most biotech start-ups' financing, the fund raising from venture capital and the capital market are not enough. Strategic alliance with big pharmaceutical or biotech companies is also necessary. At the uncertain projects in the biopharmaceuticals, the flexibility of decision-making is important against irreversibility. On the other hand, the strategic flexibility tends to become the trade-off relationship with the commitment against the preemptive behaviour by rival companies. Therefore, what kind of conditions are necessary for matching between the Pareto optimal solution on the flexibility value and the Nash equilibrium on the commitment value for the formation of partnerships? Under such research question, this study applies the option-game as the new approach, by integrating between the real options for evaluating the flexibility and the game theory for estimating the commitment. Then a research objective is to bridge the gap between a long-term technological potential and a more short-term survival, through, strategic partnerships of the biotech start-ups, whose majority is always in deficit. In conclusion, this study found some knowledge about parameter setting forward open innovation by simulation, based on option-game models.

Overview

Considering the background of both, the aging in developed nations and the basic research investment by governments in life science, the biotechnological industry is expected to be the generational industry. In particular, a biotechnological start-up has a function to facilitate the commercialization of science and technology more quickly than a big company does, but, on the other hand, has a problem of inefficient survival strategy.

This article searches for a possibility of option-game as an experimental methodology to optimize a trade-off between flexibility and commitment, as the flexibility to deal with the uncertainty and the commitment (or preemption) for a new-entrant company to compete with the predominance of existing big companies.

In other words, ROA (Real Options Analysis) is applied for analysis on the improvement of NPV (Net Present Value) in a development investment with flexibility. On the other hand, GT (Game Theory) is used for the examination on commitment taking a few of the temporary exclusive functions of competition as patents, orphan drugs, SBIRs (Small Business Innovation Researches), and CRADAs (Cooperative Research and Development Agreements). In this way, ROA examines the expanded NPV for getting over the death-valley, and GT deals with the strategic NPV by relaxation of competition with existing rivals. In short, an expectation of a certain temporary paradise, after having left a tunnel of death-valley, may promote the creation or survival of biotechnological start-ups. However, in many cases, the flexibility to wait until the decrease of uncertainty and the commitment to proactively preempt rivals, are trade-off relationships against each other. Then it is necessary to clarify strategic conditions and guidelines to decide the optimization between both functions.

Therefore, an exploratory methodology of option-game is applied to this problem. Specifically, after the explanation of a concept of option-game, the decision conditions on each proprietary or sharing strategy of developed results are examined at both quantitative and pricing competitions at duopoly, being based on the condition of a bigger NPV than WACC (Weighing Average of Cost of Capital), and at least a positive NPV. After that, from these findings, clearance conditions are searched for the strategic partnerships, for example, license or cooperative development between biotechnological start-ups and big pharmaceutical companies, or furthermore, for the open innovation.

In this way, one of the objectives is to find implications regarding strategic alliances between biotech start-ups and big pharmaceutical companies, through numerical calculations in each model of the proprietary and the sharing strategies, in the quantitative and the pricing competitions.

Application of Option-Game to Strategic Partnership

Potential of Biopharmaceuticals

In the United States of America (USA), leading the global biopharmaceuticals industry, the Federal Government's research and development (R&D) budget in the healthcare field has drastically.
increased during 1998–2004, and has separated the space field and other fields considerably.[1] Similarly, USD 29.8 billion or 54.4 percent of FY 2005 basic-research budget of the Federal Government is invested in the field of life science, mainly in medical schools, and greatly widens a difference with the second place, the engineering field of USD 9.1 billion or 16.6 percent. Since such a concentrated investment of the basic-research budget is made in a life science, it is understandable that the US Federal Government may make biotechnological industry an upbringing target for next generation industry.

The 2006 sales figures of higher-rank medicines in big pharmaceutical companies include, Pfizer's Lipitor with USD 12,890 billion at the first place, AstraZeneca's Nexium with USD 5,182 billion at the second place, and Pfizer's Norvasc with USD 4,866 billion at the third place. On the other hand, those of biotechnological companies include Amgen's Aranesp with USD 4,121 million at the first place, Amgen's Enbrel with USD 2,879 million at the second place (it has a joint sale with Wyeth, but this figure shows Amgen's part), and similarly, Amgen's Neulasta, USD 2,791 million at the third place.[2] With the sales of Lipitor being 3,127.8 times larger than Aranesp, a market difference between big pharmaceutical companies and biopharmaceutical companies is huge under the present conditions.

However, if limiting to the USA market, after three major medicines of big pharmaceutical companies — Lipitor with USD 8.6 billion at the first place, Nexium with USD 5.1 billion at the second place, and GSK's Advair Diskus with USD 3.9 billion at the third place — there are biopharmaceutical entries such as Aranesp with USD 3.9 billion at the fourth place and Amgen's Epogen with USD 3.2 billion at the sixth place. Thus, in the USA market, sales performance of biopharmaceuticals is outstanding.

And, in terms of market capitalization (cap) at the end of December 2006, USD 5.5 billion of Genentech is located in a middle ranking between USD 108.2 billion of Merck and USD 82.7 billion of Abbott Laboratories in Figure 3.1. Also, USD 79.6 billion of Amgen is located between USD 82.6 billion of AstraZeneca and USD 76.1 billion of Wyeth. Thus, the expectation of the future potential of biopharmaceuticals is large, when comparing their market cap, with the sales or profits. This fact is also reflected in higher returns of investment of main biotechnological companies than those of the many big pharmaceutical companies.

But as there are considerably many net-loss biotechnological companies, even in higher ranking market cap, for the necessary R&D costs comparable to big pharmaceutical companies, the valuation of potential for the technologies and business becomes important.

For big pharmaceutical companies, their conventional 'blockbuster model' of developing a few standard medicines for large scale and uniform markets is facing a limit. Then in a switch-over to a safe and high functional 'personal medicine model', strategic partnerships with biotechnological start-ups are possible to become the key element. Such an increase of partnership opportunity with big pharmaceutical companies can, even for biotechnological start-ups, contribute to the expansion of survival possibility, from a perspective of financial and risk management.

Figure 3.1 Big Pharmaceuticals and Main Biotechnology Companies

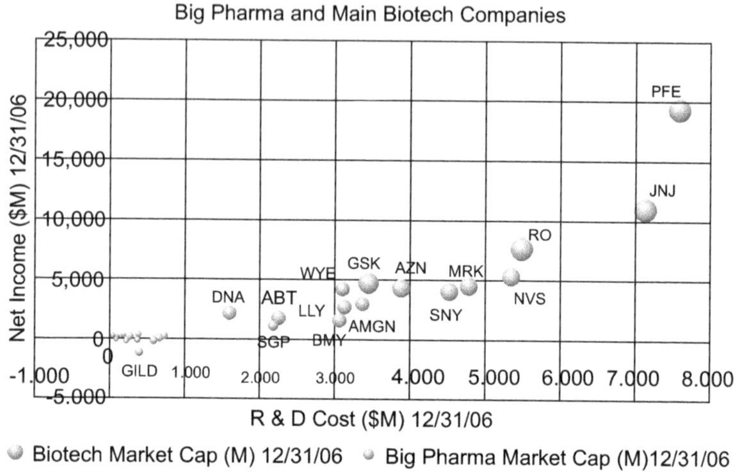

Source: On-line Wall Street Journal, 2006, http://markets wsj.com/us (accessed 31 December 2006).

Financial Conditions of Biotechnological Start-ups

Keeping in mind the continuous process of new start-ups and shut downs there are about 1,500 biotechnological start-ups, and about 300 public companies. However, since such a large burden as necessary R&D cost and long-term gap between investment and return, the number of companies in the black are limited to a few dozen within higher performance ranks. In this way, if a potential of technology and business model is evaluated, regardless of a long-term deficit, there are possibilities of financing through venture capital (VC) and Initial Public Offering (IPO). In fact, it is very mysterious that so many companies can survive even when in the red.

Certainly, in reality, there are a few biotechnology companies which are comparable to big pharmaceutical companies in terms of company value, due to high performance of the biopharmaceuticals at the product and capital markets. But most companies complementarily procure a part, equivalent to about 50 percent of the amount of money supplied from financial institutions and capital markets, by additional revenues from strategic partnerships with big pharmaceutical companies (Figure 3.2). In other words, only a company's independent financial raising from financial institutions and capital markets is not enough for many to endure during the death-valley period until the break-even point. It is worthy to mention at this point that strategic partnerships with big pharmaceutical companies as potential rivals are inevitable financial alternatives for their survival.

Figure 3.2 US Biotechnological Industry's Financial Sources

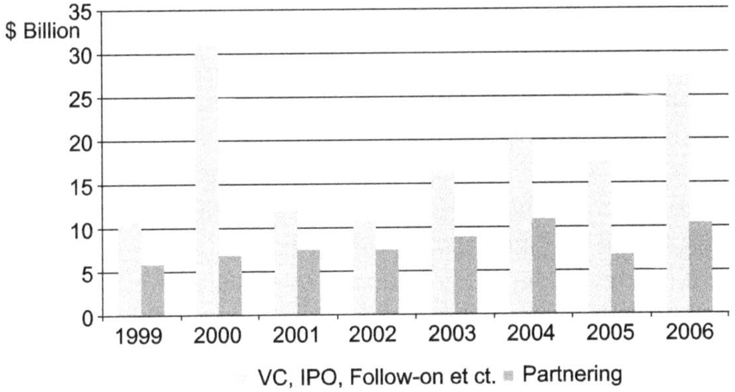

Source: Burrill (2007).

Such additional and financial trading relationships prove the existence of a pathway of R&D cooperation among universities, start-ups, and big pharmaceutical companies. In addition, license-outs are important for real options, but simultaneously they are also important elements of strategic decision-making in the option-game.

Technology Transfer Function of Biotechnological Start-up

It is the function of a biotechnological start-up to bridge over the expanding gap in the drug development process between the basic research in medical schools and the clinical development in big pharmaceutical companies. One of the big reasons for the recent shift of R&D resources from basic research to specialization in clinical development by the big pharmaceutical companies is attributed to the pressure for drug safety and pipeline evaluation. A technology-transfer function of biotechnology start-ups is to transform the basic research findings at universities, supported by a large amount of governmental research fund, into the application research more quickly and economically than big pharmaceutical companies do, by matching a radical innovation with a niche market. Thereafter the function also needs a risk management in a transition period of development, and decide whether to transfer the technology into big pharmaceutical companies or to commercialize it in its own company, if there is an opportunity.

There are advantages of technology transfer by biotechnological start-ups, as they can seemingly include the advanced function of high-expertise knowledgeable small group, the rapid adjustable small-scale company, the quick processing ability, and the low cost development. However, on the other hand, the main disadvantage of start-ups is the difficulty of continuation of development and survival. A survey states that about 60 percent of start-ups go bankrupt in six years after their foundation.[3] Under a survival strategy of the fecundity and many deaths in the resource constraints, the valuation method of technology and business becomes necessary, in order to overcome the death-valley at an initial deficit period after starting up.

Accordingly, in addition to conventional valuation of avoid potential by real options, it is also necessary to evaluate a avoid plan against rivals or for a strategic effect of partnerships, even with potential rivals.

Option-game

Here, option-game is explained as a concept of integration between real options and game theory.

Research Question

On value creation, an innovative and high-risk technology development like biotechnology start-up involves uncertainty. But to deal with it, a concept of real options is effective for a measurement of the value of flexibility as the expanded NPV created by a flexible decision-making, or a design of the decision-making to build flexibility in. On the other hand, at start-ups, commitment as the barrier to entry against big companies can create the strategic value on the unrealized and innovative ideas. At that time, game theory is a useful method to measure and ensure the strategic value. However, the flexibility related to reduction of uncertainty has often trade-off relationships with the commitment to preempt rivals.

In the context of the high weight of a strategic partnership as a fund-raising method following the financial trading, how must the strategic decision be made for the optimality on the trade-off between the flexibility and commitment from a perspective of NPV, in a decision of proprietary or sharing strategy of development result?

Definitions of main concepts and a framework

One methodology for experimentally solving such a trade-off is an idea of option-game, integrating game theory with real options. A definition of real options is the methodology as the application of concepts and techniques of financial derivatives to real assets such as a patent, a factory, etc., and so on. A measurement for value creation is the expanded NPV produced by a flexible decision-making under uncertainty. This method is applicable to a management of design, doing, and progress of decision-making for a possibility of overcoming the death-valley.

On the other hand, game theory measures the strategic value on the basis of the competitive advantage of a barrier to entry against potential rivals by the commitment reflecting the entrepreneurship of start-ups, and it is a methodology to lead the design and selection of strategies.

As a methodology here, the procedure consisted of comparing both the value of flexibility by real options and the commitment value by game theory in a game tree, and then utilizing it to the optimal strategic decision through the backward induction (Figure 3.3). In other words, here in a game tree, after dividing it into two cases, as a case of doing initial development investment, and as a base case of not investing (the benchmark for the comparison), this comparison style is applied to both quantitative and pricing competition modes.

Figure 3.3 Commitment and Flexibility

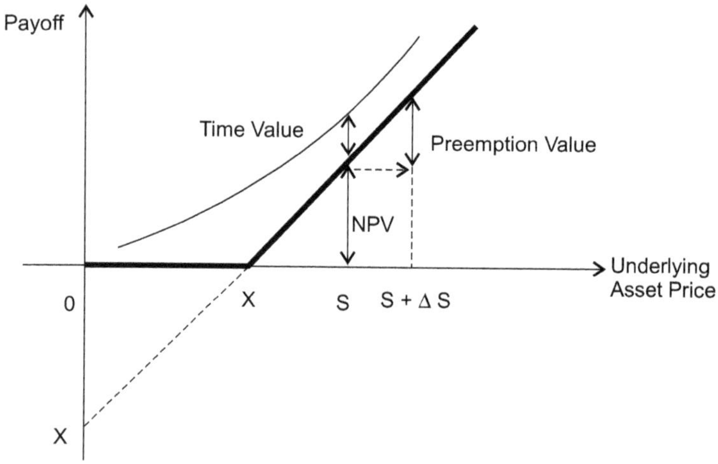

Source: Prepared by the author.

Thus, based on the presumption of duopoly competition form in game theory, at each quantitative and pricing competition, the objective is to find a possibility of positive NPV for proprietary and sharing strategies of development findings by an initial investing company or player, and to try to make clear the feasible criteria for strategic partnerships (license and cooperative development) between biotechnological start-ups and big pharmaceuticals companies, and further the opportunities for open innovation.

At a quantitative competition, a base case is defined as not having initial development investment but making commercialization investment latterly, and is based on a binomial process of up-and-down demand events. And each NPV is compared in the Cournot-Nash equilibrium, Stackelberg leader/follower equilibrium, and monopoly. Likewise, at a base case of binomial process in pricing competition, comparisons of each NPV are made among the Bertrand Nash equilibrium, Stackelberg leader/follower equilibrium, and monopoly. Next, about each competition form, after explanation on a base case as being not accompanied with such a preceding investment, a comparison of NPVs between two players is made by a proprietary or sharing strategy of this initial development investor at quantitative and pricing competition.

Literature Survey

In the preceding research works related to this study, from the result of a simplified method of approximation of the Black-Sholes equation by a binomial process, more intuitive understanding and practical application was endorsed.[4]

On real options as the application of an idea of option in financial derivatives to real assets, at first, the very idea of real options is originated in the implicit presentation when an application of finance theory was suggested for strategic planning.[5] Then, a qualitative explanation on the flexibility value as core performance of real options,[6] and a framework of the flexibility, synergy effect, and control of value maximization in capital budgeting[7] were given. Otherwise, as specific techniques, a value calculation of deferment option[8] and applied respect in abandonment option[9] were discussed.

Game theoretical analysis of investment strategy for rival entries[10] and mathematical explanation on preemptive value as analogy of commitment with dividends in options with flexibility[11] were tried.

As research examples of option-game as a new concept of more systematically integrating real options with game theory, at first, the idea of option-game was created, when the relationships between proprietary/sharing competitions and the timing of growth options were examined.[12] In addition, with a game tree analysis, the optimal investment timing by comparison between the preemptive value of commitment and the learning effect of flexibility was studied.[13]

In the meantime, by formally integrating real options with game theory, an analysis by game trees was studied.[14] Furthermore, the flexibility and commitment was formalized as option-game integrated between real options and game theory.[15]

However, existing studies on option-game have concentrated only on discrete analysis, and may have a little arbitrariness in conclusions. Therefore, a uniqueness of this paper supposes the continual interaction with parameters, and makes it a selecting guideline of the optimal path in a game tree. In addition, objectivity of a conclusion may also be improved.

Research Objective

For economic and risk evaluations of strategies of technological and business development for overcoming the death-valley in order to raise funds from financial institutions and capital markets such as VC or IPO, the effectiveness of real options is proved in a certain range. Therefore, the objectives are, firstly, to apply a concept and techniques of option-game tentatively, by reflecting such trade-off relationship to occur when having to consider a new strategic factor of the commitment concerned with the competition, and partnership with rivals in addition to the flexibility of real options, and secondarily to examine the logic and technique on rationality of decision making about optimal planning on proprietary or sharing strategy of development findings, when being assumed in strategic partnerships.

The expected results from this research are to examine results of numerical analysis on the base case, proprietary strategy and sharing strategy on the basis of specification parameters in the quantitative and pricing competitions, and then to investigate a direction toward a possibility of open innovation.

Partnership strategy in quantitative competition

Here, each result of numerical calculation is examined of the base case, proprietary strategy, and sharing strategy of development findings in the quantitative competition.

Demand and volatility of base case in quantitative Competition

At first, own positioning policy for maximizing NPV becomes possible to use, by the optimal arrangement of each decision type (Type A, Type B and Type C) among the commitment priority, the flexibility priority, and the balance between them, on the basis of demand and volatility parameters.

Here, it is assumed a two-player game between player i and j (or A and B), initial demand $\theta_0 = 17.5$, up or down with binomial parameter $u = 1.25$ and $d = 0.8$, each player's cost $c_i = 5$ and $c_j = 5$, initial R&D investment $I_0 = 100$, follow-up commercialization investment $I_1 = 100$, risk-adjusted discount rate $k = 0.13$, and risk-free interest rate $r_f = 0.1$. If constant asset payout yield for perpetual project is

$$\delta = \frac{k}{1+k}$$

risk neutral probability is

$$p = \frac{(1+r-\delta)-d}{u-d}$$
$$= 0.41101\cdots$$
$$\approx 0.411.$$

And if re-combination at binomial process is assumed,

$$u = \exp(\sigma\sqrt{\Delta t}), d = \exp(-\sigma\sqrt{\Delta t}), \text{ and volatility } \sigma.$$

In base case at second stage, general equation of Cournot-Nash equilibrium (C) is

$$\text{NPV}_i(C) = \frac{(\theta_t - 2\times c_i + c_j)^2}{9k} - I_1.$$

For example, on Cournot-Nash equilibrium at second stage, when after i selects $I_0 = 100$, demand increases twice, both i and j select D (Defer) first and I (Invest) next, and a Cournot-Nash equilibrium is attained, Net Present Value of i (NPV_i) is

$$\text{NPV}_i(C) = \text{NPV}_i(uDuI, uDuI)$$
$$= \frac{(\theta_0 \times u^2 - 2\times c_A + c_B)^2}{9k} - I_1$$
$$= \frac{(17.5\times 1.25^2 - 2\times 5 + 5)^2}{9\times 0.13} - 100$$
$$= 326.70355\cdots$$
$$\approx 327.$$

At second stage, general equation of monopoly (M) is

$$\mathrm{NPV}_i(M) = \frac{(\theta_t - c_i)^2}{4k} - I_1.$$

For example, a specific calculation case is

$$\begin{aligned}\mathrm{NPV}_i(M) &= \mathrm{NPV}_i(uDuI,\ uDuD)\\ &= \frac{(\theta_0 \times u^2 - c_A)^2}{4k} - I_1\\ &= \frac{(17.5 \times 1.25^2 - 5)^2}{4 \times 0.13} - 100\\ &= 860.08300\cdots\\ &\approx 860.\end{aligned}$$

At second stage, general equation of Stackelberg leader equilibrium (SL) is

$$\mathrm{NPV}_i(SL) = \frac{(\theta_t - 2c_i + c_j)^2}{8k} - I_1.$$

And, for example,

$$\begin{aligned}\mathrm{NPV}_i\,(SL) &= \mathrm{NPV}_i(uIuI,\ uDuI)\\ &= \frac{(\theta_0 \times u^2 - 2c_A + c_B)^2}{8k} - I_1\\ &= \frac{(17.5 \times 1.25^2 - 2 \times 5 + 5)^2}{8 \times 0.13} - 100\\ &= 380.04150\cdots\\ &\approx 380.\end{aligned}$$

At second stage, general equation of Stackelberg follower equilibrium (SF) is

$$\mathrm{NPV}_i(SF) = \frac{(\theta_t - 3c_j + 2c_i)^2}{16k} - I_1.$$

And,

$$\begin{aligned}\mathrm{NPV}_i(SF) &= \mathrm{NPV}_i(uDdI,\ uIdI)\\ &= \frac{(\theta_0 - 3c_B + 2c_A)^2}{16k} - I_1\\ &= \frac{(17.5 - 3\times 5 + 2\times 5)^2}{16\times 0.13} - 100\\ &= -24.87980\cdots\\ &\approx -25.\end{aligned}$$

At second stage, general equation of abandonment (A) is

$$\mathrm{NPV}_i(A) = 0.$$

And,

$$\begin{aligned}\mathrm{NPV}_i(A) &= \mathrm{NPV}_i(uDuD,\ uIuI)\\ &= 0\end{aligned}$$

By sharing the same equation with second stage, at first stage general equation of Cournot-Nash equilibrium (C) is

$$\mathrm{NPV}_i(C) = \frac{(\theta_t - 2\times c_i + c_j)^2}{9k} - I_1.$$

For example,

$$\begin{aligned}\mathrm{NPV}_i(C) &= \mathrm{NPV}_i(uI,\ uI)\\ &= \frac{(\theta_0\times u - 2\times c_A + c_B)^2}{9k} - I_1\\ &= \frac{(17.5\times 1.25 - 2\times 5 + 5)^2}{9\times 0.13} - 100\\ &= 143.38942\cdots\\ &\approx 143.\end{aligned}$$

At first stage, general equation of monopoly (M) is

$$\mathrm{NPV}_i(M) = \frac{p\times V_u + (1-p)\times V_d}{1+r_f} - I_1 + \frac{\pi M}{1+k}.$$

where,

πM = monopolistic profit limited only at first stage.

For instance,

$$\begin{aligned}
\text{NPV}_i(M) &= \text{NPV}(uI, uD) \\
&= \frac{p \times V_u + (1-p) \times V_d}{1+r_f} - I_1 + \frac{(\theta_t \times u - c_i)^2}{4k} \times \frac{1}{1+k} \\
&= \frac{0.411 \times 380 + (1-0.411) \times 200}{1+0.1} - 100 \\
&\quad + \frac{(17.5 \times 1.25 - 5)^2}{4 \times 0.13} \times \frac{1}{1+0.13} \\
&= 303.25808 \cdots \\
&\approx 303.
\end{aligned}$$

At first stage, general equation of deferment is

$$\text{NPV}_i(D) = \frac{p \times NPV_u + (1-p)NPV_d}{1+r_f}.$$

For instance,

$$\begin{aligned}
\text{NPV}_i(D) &= \text{NPV}_i(uD, uI) \\
&= \frac{p \times NPV_u + (1-p)NPV_d}{1+r_f} \\
&= \frac{0.411 \times 140 + (1-0.411) \times 0}{1+0.1} \\
&= 52.31847 \cdots \\
&\approx 52.
\end{aligned}$$

Next, from a 3-dimensions graph by the demand, volatility, and NPV, such a trade-off relationship is understandable that the commitment is emphasized for demand and the flexibility is made much for volatility (Figure 3.4). In addition, the optimal strategy can be chosen with each parameter from a viewpoint of option-game, from Table 3.1 as a map (for a more detailed strategy, see

Type A in Figure 3.5, Type B in Figure 3.6 and Type C in Figure 3.7, respectively).

From an option-game perspective, the optimal strategic combination with a rival is located on this 3-dimensions graph, if such 2 parameters are preliminarily decided.

Figure 3.4 Base Case of Qualitative Competition

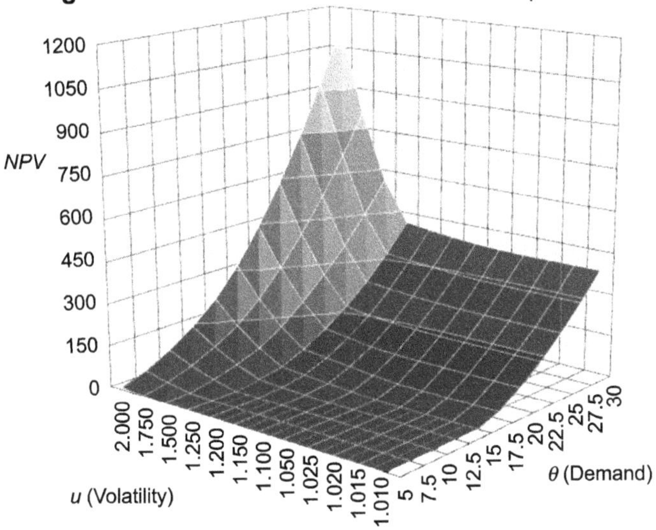

Source: Prepared by the author.

Table 3.1 Three Types in Base Case of Quantitative Competition

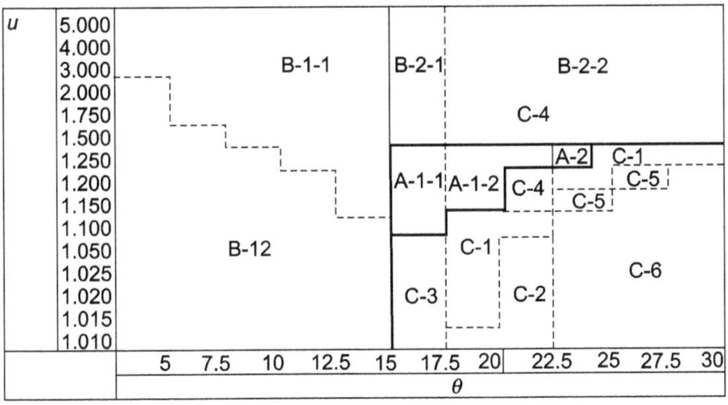

$C_i=C_j=5$

Source: Prepared by the author.

Figure 3.5 Type-A Tree

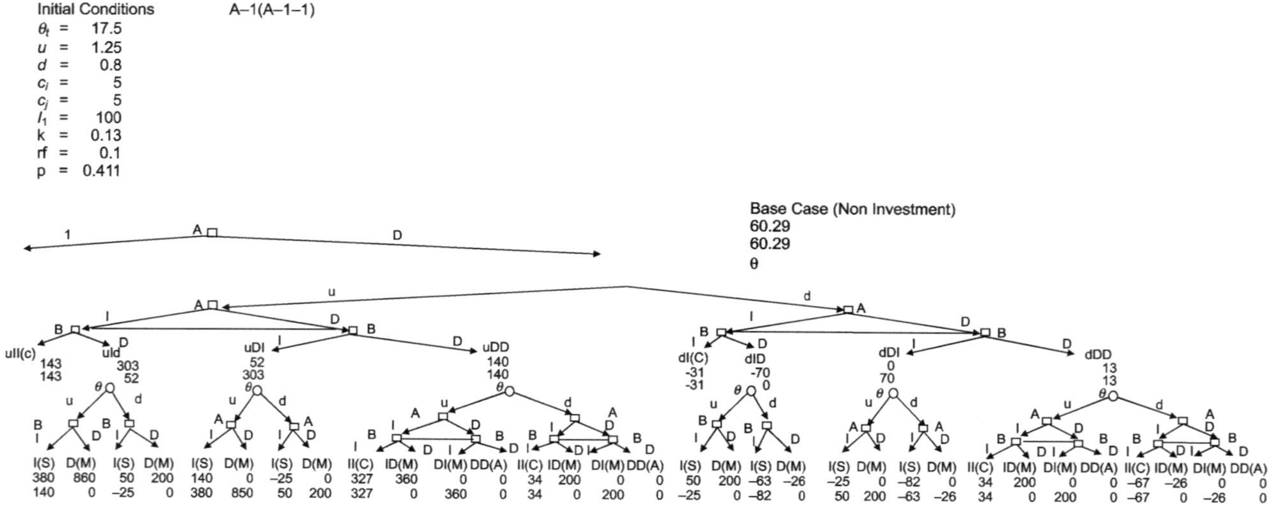

Source: Prepared by the author.

Figure 3.6 Type-B Tree

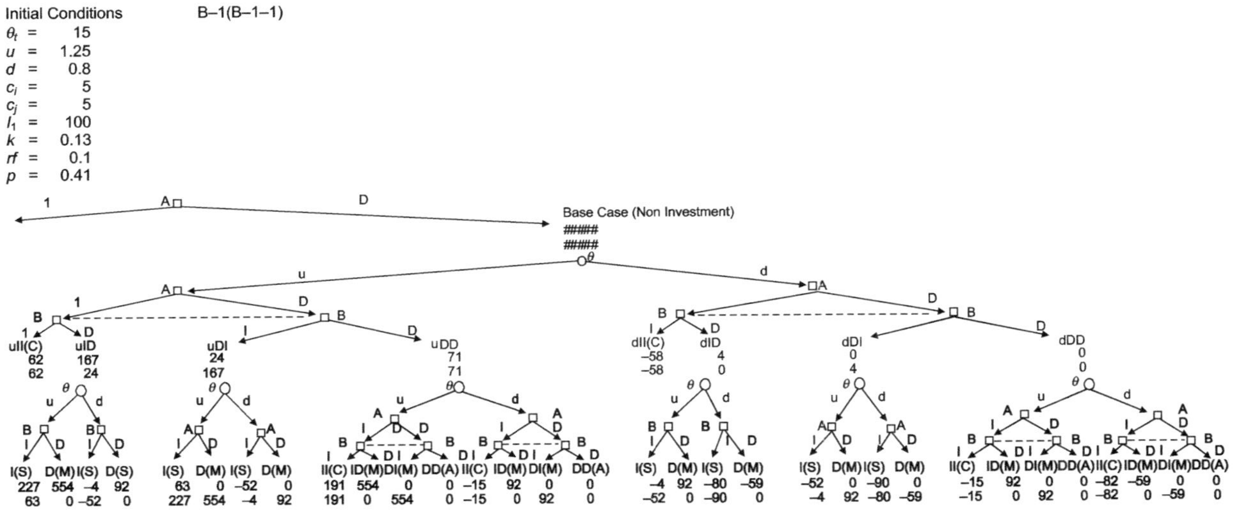

Source: Prepared by the author.

Figure 3.7 Type-C Tree

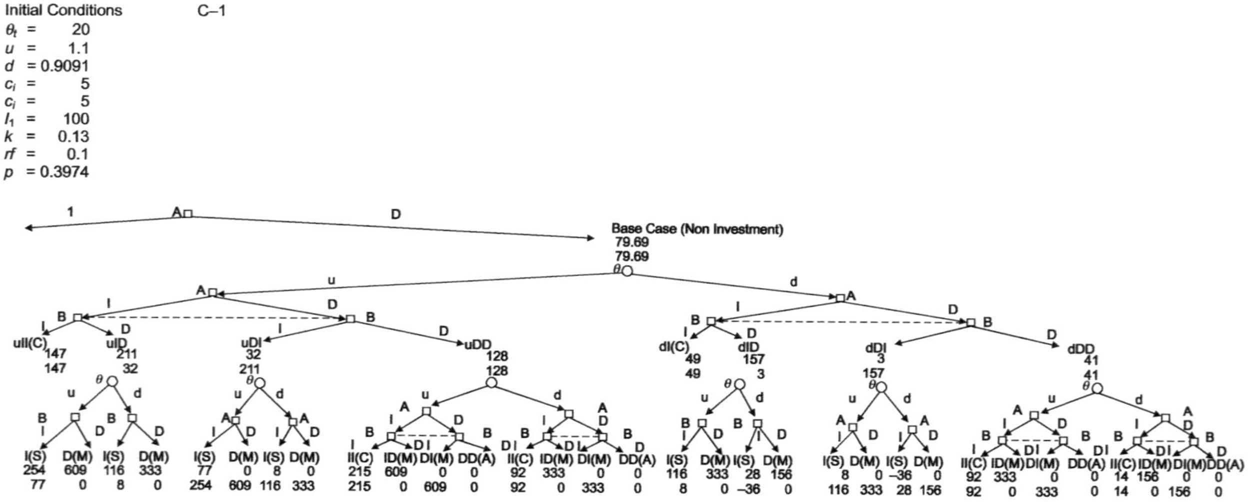

Source: Prepared by the author.

Proprietary Strategy of R&D Findings in Quantitative Competition

Advantage of Pioneering company with cost-reduction effect

The advantage of proprietary strategy of R&D findings in quantitative competition could be shown by the achievement degree of cost reduction.

At first, a pattern is examined for a company investing in initial R&D to find a predominant position of NPV against a rival company with no initial investment, and against a base case (in the case of no initial R&D investing), in a way, with the achievement degree of cost reduction (Figures 3.8 and 3.9). As a result, a rival company without initial investment tends to decrease the NPV with expansion of the cost effect of a pioneering company as an initial R&D investor.

In this way, the bigger the cost-reduction effect is, the more the advantage against a rival company.

Figure 3.8 Impact of Cost Saving by Proprietary Strategy in Quantitative Competition

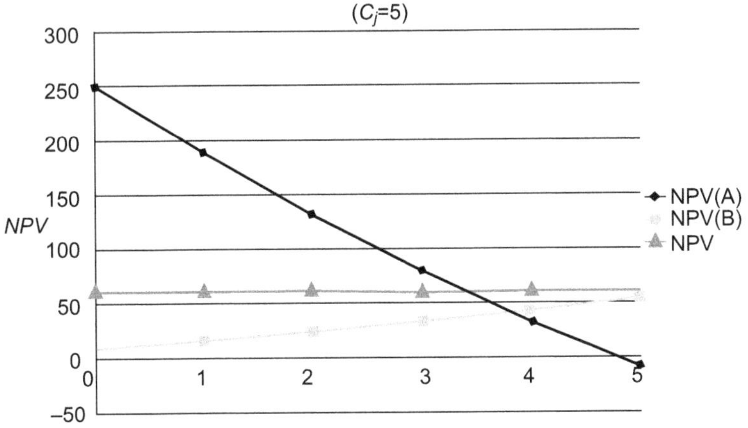

Source: Prepared by the author.

Figure 3.9 Proprietary Strategy Tree in Quantitative Competition

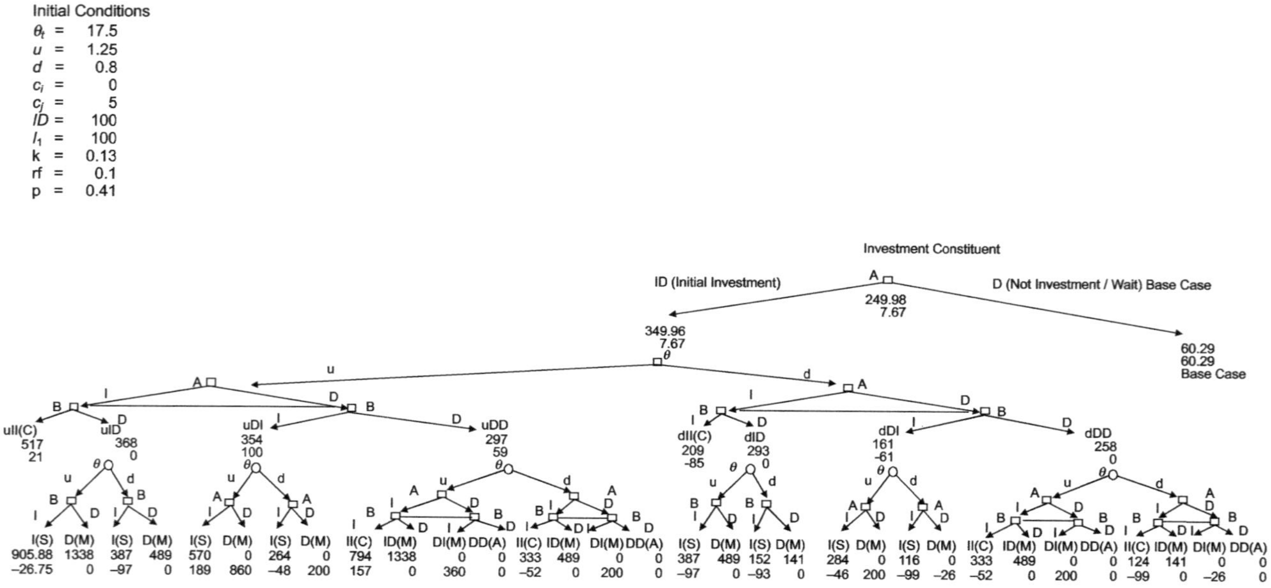

Source: Prepared by the author.

Impact of Demand

The influence of a demand on NPV is investigated, by a standard of cross points with a rival company in cost-reduction effect (Figure 3.10). As a result, there is a situation that in the area of more than a standard demand (17.5), some advantage of pioneering company is acquired, and also there is a tendency that such advantage is lost by the demand reduction from a standard. However, the curves of both the companies show an inclination to move, while considerably coming close in the midway of a base case.

Figure 3.10 Impact of Demand Change by Proprietary Strategy in Quantitative Competition

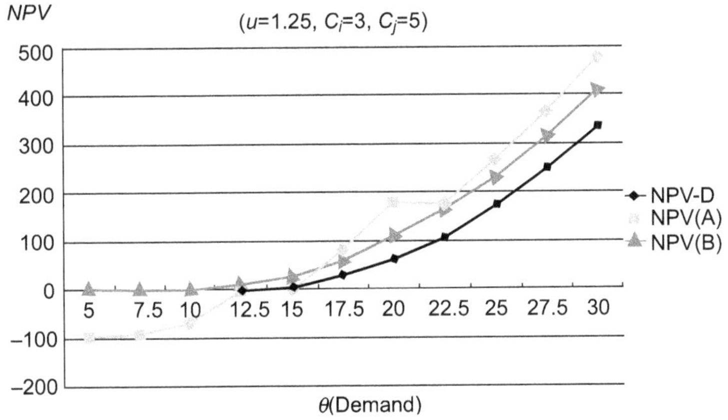

Source: Prepared by the author.

Impact of Volatility

In the area of lower volatility after a cross point (u =1.25), the advantage of a pioneering company's NPV is constantly maintained (Figure 3.11). On the other hand, in the area of higher volatility, the advantage of pioneering company recovers, except a mid-range of advantageous area of the rival company without initial investment.

Sharing Strategy of R&D Findings in Quantitative Competition

When there is no means to charge an investment benefit, and a pioneering company with initial R&D investment, shares development findings with a rival company, if conditions are constant, the pioneering company's position becomes disadvantageous, when compared with a base case.

Figure 3.11 Impact of Volatility Change by Proprietary Strategy in Quantitative Competition

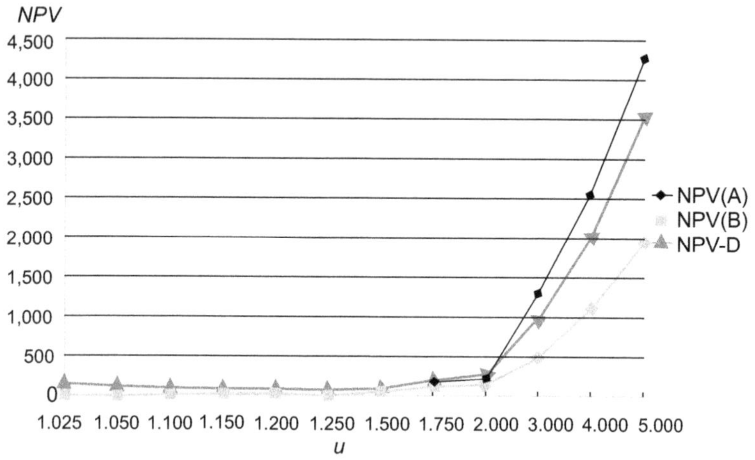

Source: Prepared by the author.

However, it is possible to attain to the same level of a base case, by a certain combination of demand and volatility, and even possible to set a condition to exceed the level of a base case. In addition, if the pioneering side not only always pays attention to disadvantageous condition but also estimates the positive effects for both sides, it is possible for the pioneer to draw a policy to collect profits from a free rider enjoying profits without initial investment.

Partnership Strategy in Pricing Competition

Here, each numerical calculation on a base case, a proprietary strategy, and a sharing strategy of R&D findings is shown (Figure 3.12).

Figure 3.12 Sharing Strategy in Quantitative Competition

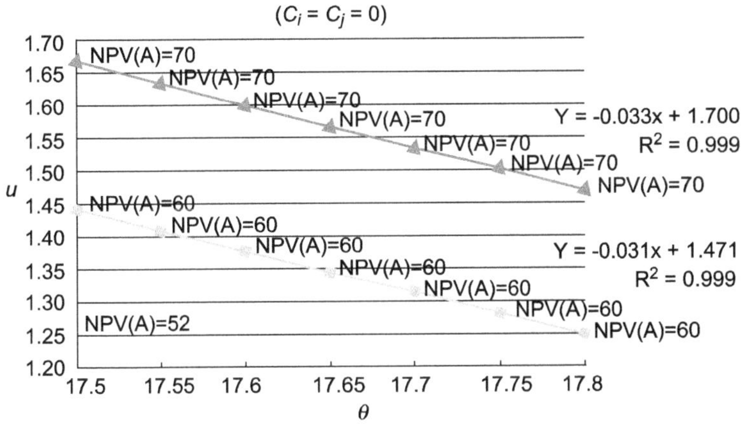

Source: Prepared by the author.

Demand and volatility of base case in pricing competition

Areas of each type, A, B, and C on demand and volatility occur likewise the same image with the case of quantitative competition. The optimal strategy for its own NPV with a rival company is fixed with this area map. Type C is equivalent to commitment, type B is to flexibility, and type A is to an intermediate region of both types.

As parameter conditions belonging to type A, there are initial demands $\theta_{i0} = 12$, $\theta_{j0} = 12$, up rate u = 12.5, down rate $d = 0.8$, price sensitivity coefficients $b = 2$, $dd = 1$, costs $c_i = 3$, $c_j = 3$, initial development investment $I_0 = 100$, commercialization investment $I_i = 100$, risk adjusted discount rate $k = 0.13$, risk free interest rate $r_f = 0.1$, and risk neutral probability as the same condition with quantitative completion $p = 0.411$.

General equation of Bertrand Nash equilibrium in common with first and second stages is

$$\mathrm{NPV}_i(N) = \frac{(PN_i - c_i)(\theta_{it} - b \times PN_i + dd \times PN_j)}{k} - I_1.$$

As an example of Bertrand Nash equilibrium at second stage

$$\mathrm{NPV}_i(N) = \mathrm{NPV}_i(uDuI,\ uDuI)$$

$$= \frac{\begin{matrix}(u^2 \times PN_A - c_A)\\(\theta_{A0} \times u^2 - b \times u^2 \times PN_A + dd \times u^2 \times PN_B)\end{matrix}}{k} - I_1$$

$$= \frac{\begin{matrix}(1.25^2 \times 6 - 3)\\(12 \times 1.25^2 - 2 \times 1.25^2 \times 6 + 1 \times 1.25^2 \times 6)\end{matrix}}{0.13} - 100$$

$$= 324.03846 \cdots$$

$$\approx 324.$$

The Nash equilibrium price is

$$PN_i = \frac{\{2b(\theta_{it} + b \times c_i) + dd(\theta_{jt} + b \times c_j)\}}{4b^2 - dd^2}$$

$$= \frac{\{2b(\theta_{A0} + b \times c_A) + dd(\theta_{B0} + b \times c_B)\}}{4b^2 - dd^2}$$

$$= \frac{\{2 \times 2(12 + 2 \times 3) + dd(12 + 2 \times 3)\}}{4 \times 2^2 - 1^2}$$

$$= 6.$$

General equilibrium of monopoly at second stage is

$$\mathrm{NPV}_i(M) = \frac{(PM_i - c_i)\theta_{it}}{k} - I_1.$$

As an example of monopoly of second stage,

$$\mathrm{NPV}_i(M) = \mathrm{NPV}_i(uIuI,\ uDuD)$$

$$= \frac{(u^2 PM_A - c_A)\theta_{A0} u^2}{k} - I_1$$

$$= \frac{(1.25^2 \times 7.5 - 3)12 \times 1.25^2}{0.13} - 100$$

$$= 1035.81730 \cdots$$

$$\approx 1036.$$

The monopoly price is

$$PM_i = \frac{\theta_{it} + c_i(b - dd)}{2(b - dd)}$$

$$= \frac{\theta_{A0} + c_A(b - dd)}{2(b - dd)}$$

$$= \frac{12 + 3(2 - 1)}{2(2 - 1)}$$

$$= 7.5$$

At second stage, Stackelberg leader equilibrium equation is

$$\text{NPV}_i(SL) = \frac{(PSL_i - c_i)(\theta_{it} - b \times PSL_i + dd \times PSF_j)}{k} - I_1.$$

Stackelberg follower equilibrium equation at second stage is

$$\text{NPV}_i(SF) = \frac{(PSF_i - c_i)(\theta_{it} - b \times PSF_i + dd \times PSL_j)}{k} - I_1.$$

As an example of Stackelberg leader equilibrium at second stage,

$$\text{NPV}_i(SL) = \text{NPV}_i(uIuI,\ uDuI)$$

$$= \frac{(u^2 \times PSL_i - c_i)(\theta_{it} \times u^2 - b \times u^2 \times PSL_i + dd \times u^2 \times PSF_j)}{k} - I_1$$

$$= \frac{(u^2 \times PSL_A - c_A)(\theta_{A2} \times u^2 - b \times u^2 \times PSL_A + dd \times u^2 \times PSF_B)}{k} - I_1$$

$$= \frac{(1.25^2 \times 6.2 - 3)(12 \times 1.25^2 - 2 \times 1.25^2 \times 6.2 + 1 \times 1.25^2 \times 5.8)}{0.13} - 100$$

$$= 315.26871 \cdots$$

$$\approx 315.$$

As an example of Stackelberg follower equilibrium at second stage, although the value is larger than Stackelberg leader equilibrium due to price sensitivity coefficient,

$$\text{NPV}_i(SF) = \text{NPV}_i(uDuI, uIuI)$$

$$= \frac{(u^2 \times PSF_i - c_i)(\theta_{it} \times u^2 - b \times u^2 \times PSF_i + dd \times u^2 \times PSL_j)}{k} - I_1$$

$$= \frac{(u^2 \times PSF_A - c_A)(\theta_{A2} \times u^2 - b \times u^2 \times PSF_A + dd \times u^2 \times PSL_B)}{k} - I_1$$

$$= \frac{(1.25^2 \times 5.8 - 3)(12 \times 1.25^2 - 2 \times 1.25^2 \times 5.8 + 1 \times 1.25^2 \times 6.2)}{0.13} - 100$$

$$= 338.38364 \cdots$$

$$\approx 338.$$

The price of Stackelberg leader equilibrium is

$$PSL_i = \frac{2b(\theta_{it} + b \times c_i) + dd(\theta_{jt} + b \times c_j - dd \times c_i)}{4b^2 - 2dd^2}$$

$$= \frac{2 \times 2(12 + 2 \times 3) + 1(12 + 2 \times 3 - 1 \times 3)}{4 \times 2^2 - 2 \times 1^2}$$

$$= 6.21428 \cdots$$

$$\approx 6.2$$

The price of Stackelberg follower equilibrium is

$$PSF_j = \frac{2b \times dd \times \theta_{it} + dd(2b - dd^2)c_i + (2b + dd)(2b - dd)(\theta_{jt} + b \times c_j)}{4b(2b^2 - dd^2)}$$

$$= \frac{2\times 2\times 1\times 12+1(2\times 2-1^2)\times 3+(2\times 2+1)(2\times 2-1)(12+2\times 3)}{4\times 2(2\times 2^2-1^2)}$$

$$= 5.83928\cdots$$
$$\approx 5.8$$

General equation of abandonment at second stage is the same one with quantitative competition,

$$\text{NPV}_i(A) = 0.$$

For example,

$$\text{NPV}_i(A) = \text{NPV}_i(uDuD, uIuI)$$
$$= 0.$$

At first stage, general equation of Bertrand Nash equilibrium is the same one with first stage,

$$\text{NPV}_i(N) = \frac{(PN_i - c_i)(\theta_{it} - b\times PN_i + dd\times PN_j)}{k} - I_1.$$

For example,

$$\text{NPV}_i(N) = \text{NPV}_i(uI, uI)$$

$$= \frac{(uPN_i - c_i)(\theta_{it} - b\times uPN_i + dd\times uPN_j)}{k} - I_1$$

$$= \frac{(uPN_A - c_i)(\theta_{A1} - b\times uPN_A + dd\times uPN_B)}{k} - I_1$$

$$= \frac{(uPN_A - c_i)(\theta_{A1} - b\times uPN_A + dd\times uPN_B)}{k} - I_1$$

$$= \frac{(1.25\times 6-3)(12-2\times 1.25\times 6+1\times 1.25\times 6)}{0.13} - 100$$

$$= 146.15384\cdots$$

$$= 146.$$

At first stage, general equation of monopoly is

$$\text{NPV}_i(M) = \frac{p\times V_u + (1-p)\times V_d}{1+r_f} - I_1 + \frac{\pi M}{1+k}.$$

where,
πM = monopolistic profits only at first stage.
For example,

$$\text{NPV}_i(M) = \text{NPV}_i(uI, uD)$$

$$= \frac{p(V_u + I_1) + (1-p)(V_d + I_1)}{1+r_f} - I_1 + \frac{(u \times PM_i - c_i)\theta_{it} \times u}{1+k}$$

$$= \frac{p(V_u + I_1) + (1-p)(V_d + I_1)}{1+r_f} - I_1 + \frac{(u \times PM_A - c_A)\theta_{A1} \times u}{1+k}$$

$$= \frac{0.411(314+100) + (1-0.411)(34+100)}{1+0.1} - 100 + \frac{(1.25 \times 7.5 - 5)12 \times 1.25}{1+0.13}$$

$$= 206.44263 \cdots$$

$$\approx 206.$$

At first stage, general equation of deferment is

$$\text{NPV}_i(D) = \frac{p \times NPV_u + (1-p)NPV_d}{1+r_f}.$$

For instance,

$$\text{NPV}_i(D) = \text{NPV}_i(uD, uI)$$

$$= \frac{p \times NPV_u + (1-p)NPV_d}{1+r_f}$$

$$= \frac{0.411 \times 338 + (1-0.411)43}{1+0.1}$$

$$= 149.32350 \cdots$$

$$\approx 149.$$

When a 3-dimension map is used, it is possible to find the optimal NPV level by both demand and volatility (Figure 3.13). Roughly, type C showing commitment is selected in the demand-oriented area, type B showing flexibility is chosen in a volatility-oriented area, and type A is preferred in a central area between them (for the total map, see Table 3.2 and for an example of Type A, see Figure 3.14).

Figure 3.13 Base Case in Price Competition

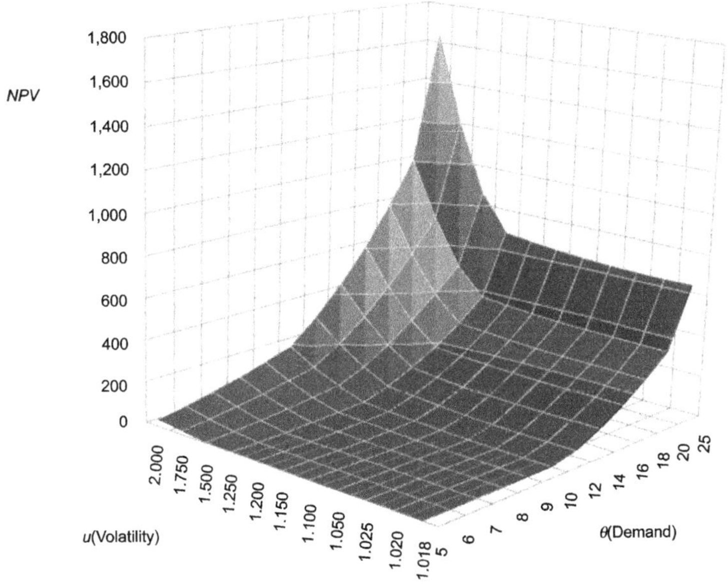

Source: Prepared by the author.

Table 3.2 Tree Types in Base Case of Price Competition

		Q												
		5	*6*	*7*	*8*	*9*	*10*	*12*	*14*	*16*	*18*	*20*	*25*	*30*
	5	B	B	B	B	B	B	B	B	B	B	B	B	B
	4	B	B	B	B	B	B	B	B	B	B	B	B	B
	3	B	B	B	B	B	B	B	B	B	B	B	B	B
	2	B	B	B	B	B	B	B	B	B	B	B	B	B
	1.75	B	B	B	B	B	B	B	B	B	B	B	B	B
	1.5	B	B	B	B	B	B	B	B	B	B	B	B	B
u	1.25	B	B	B	B	B	B	A	A	A	C	C	C	C
	1.2	B	B	B	B	B	B	A	A	C	C	C	C	C
	1.15	B	B	B	B	B	B	A	C	C	C	C	C	C
	1.1	B	B	B	B	B	B	A	C	C	C	C	C	C
	1.05	B	B	B	B	B	B	C	C	C	C	C	C	C
	1.025	B	B	B	B	B	B	C	C	C	C	C	C	C
	1.02	B	B	B	B	B	B	C	C	C	C	C	C	C
	1.0175	B	B	B	B	B	B	C	C	C	C	C	C	C

Source: Prepared by the author.

Figure 3.14 Type-A Tree of Base Case in Price Competition

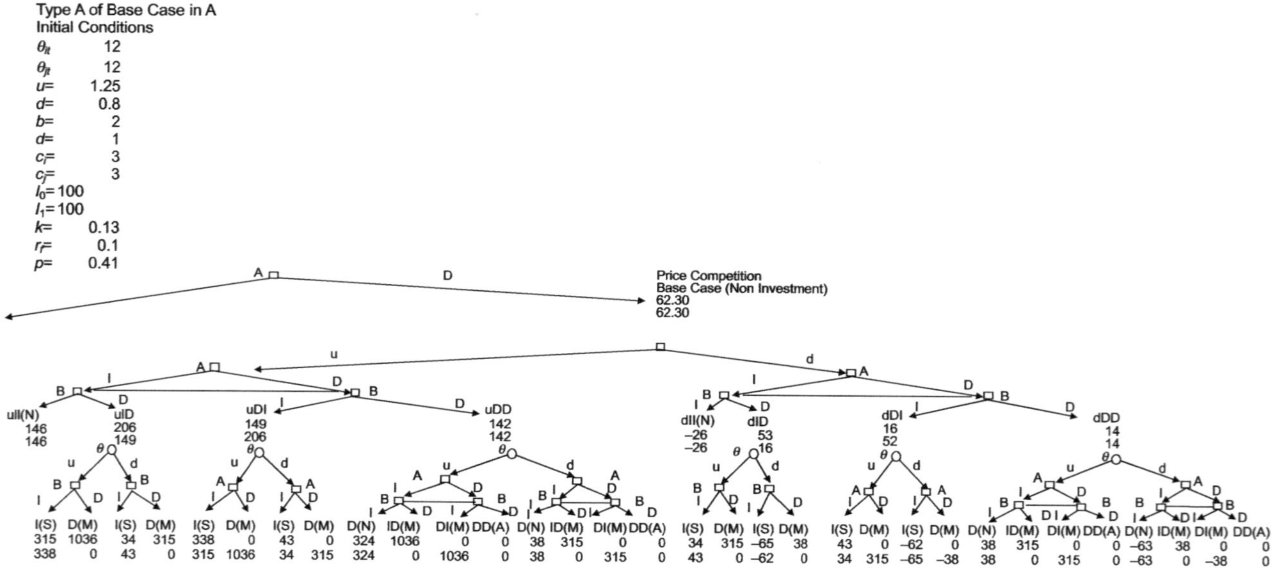

Source: Prepared by the author.

Proprietary Strategy of R&D Findings in Pricing Competition

Impact of demand

During a range of 9-20 of pioneering company's demand, tree type is basically type C except at demand level 12, and accompanying small changes in transient period of 10–13. In Bertrand Nash equilibrium at price competition, an equilibrium point of zero profit is theoretically the end point. However, if market segmentation is produced by product differentiation, niche segregation is possible, and this example assumes such a state.

When the demand of pioneering investing company A is expanding, while the demand of non-investing company B is fixed, it shows a possibility that company A overcomes the disadvantage of initial independent investing, and surpasses the NPV of a base case and even that of a rival company (Figure 3.15). This displays the area possible of net positive return for initial R&D investing.

Figure 3.15 Impact of Demand Change by Proprietary Strategy in Price Competition

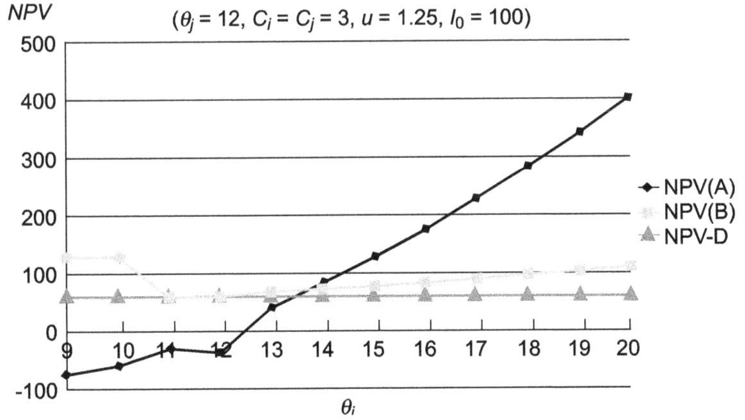

Source: Prepared by the author.

Impact of volatility

When volatility changed in 1.5 from 1.25, a form of almost the same cross of curves of both A–B companies is shifting in a direction on the right being cross in a range of 9–20 demand (Figure 3.16). This is regarded as an effect of volatility. While demand is less than 11, it shows a pattern of zero-sum. But when demand is more than 12, a synergy effect is seen.

Figure 3.16 Impact of Volatility Change by Proprietary Strategy in Price Competition

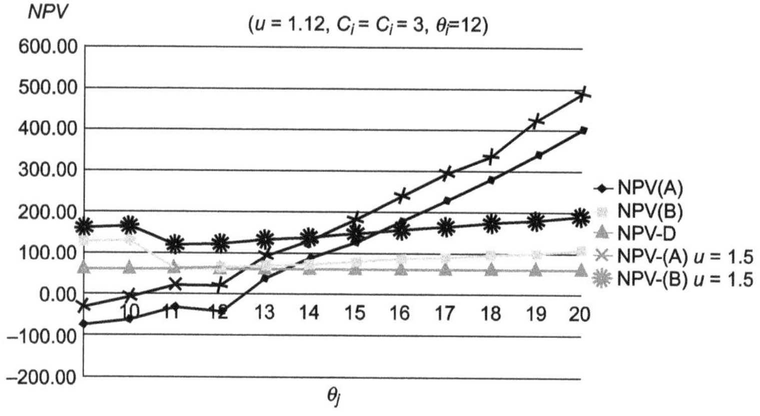

Source: Prepared by the author.

Impact of cost reduction

When cost is reduced from 3 to 0 by initial R&D investment, during a demand range of 12–20, there is an effect of a rising shift of its NPV for a pioneering company A, while maintaining a slope of the curve (Figure 3.17). Otherwise, for non-investing company B, there is an effect of shift down for NPV, while maintaining a rise tendency of the curve slope. Besides, the width of a curve rise in company A is bigger than the width of a curve drop in company B. In this way, a cost-reduction effect is considerably profitable to a pioneering company, but, is slightly disadvantageous for a non-investing company.

Figure 3.17 Impact of Cost Saving by Proprietary Strategy in Price Competition

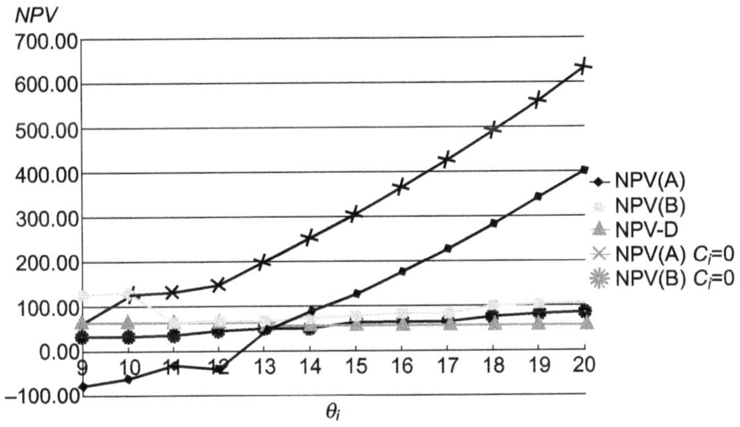

Source: Prepared by the author.

In price competition, an advantage of cost reduction may greatly have an influence on a market expansion. Therefore, when demand is more than 12 of a standard, the advantage of acquisition in market expansion is reflected, in a big rise shift for pioneering company A, and a small drop shift for non-investing company B. Furthermore, even when demand is less than a standard in 9–11, an advantage of market share is seen by cost reduction of a pioneering company.

Sharing strategy of R&D findings in Pricing Competition

Even if a common demand change between company A and B is assumed, the inferior level of a pioneering investing company A is understandable from the difficulty of investment return (Figure 3.18). But if demand becomes large, even on the assumption of inferior level of company A, it is possible to change that level of NPV higher level than a base case. The curve of company B is moving in parallel with the curve of company A, by maintaining the advantage of non-investing initial R&D cost. Company A converts a minus NPV into positive one in a demand level of common 13–14, and realizes higher NPV than a base case in a level of 15–16. Also, in the case of cost reduction, it is possible to change a higher level of NPV of company A than a base case, while leaving inferior level of company A.

As a 3-dimension graph shows a change of NPV by the demand and volatility as common to company A and B, an improvement of NPV can be seen, reflected in the trade-off between demand and volatility (Figure 3.19).

Figure 3.18 Demand Change at Sharing Strategy in Price Competition

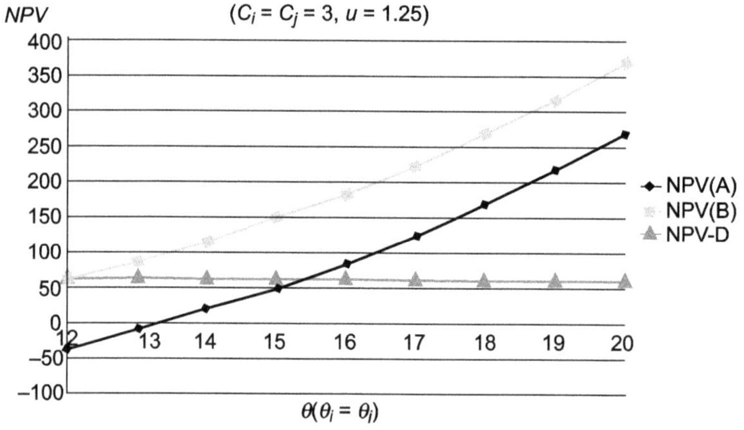

Source: Prepared by the author.

Thus, in both quantitative and pricing competitions, a sharing strategy of R&D findings shows a possibility of working disadvantageously for a pioneering investment company. But by a parameter setting of demand and volatility, it is possible to improve that NPV rather than a base case. In addition, in Pareto optimum as a result of both companies total, NPV of a sharing strategy is higher than that of a proprietary strategy. Then it is possible to attain to plus-sum by preliminarily preventing a free-riding of a late-departure company, and by adopting a sharing strategy of R&D findings. After taking into account a method to recover some investment return from benefit enjoyment, a pioneering company should decide on a condition to share results.

Figure 3.19 Demand and Volatility Changes at Sharing Strategy in Price Competition

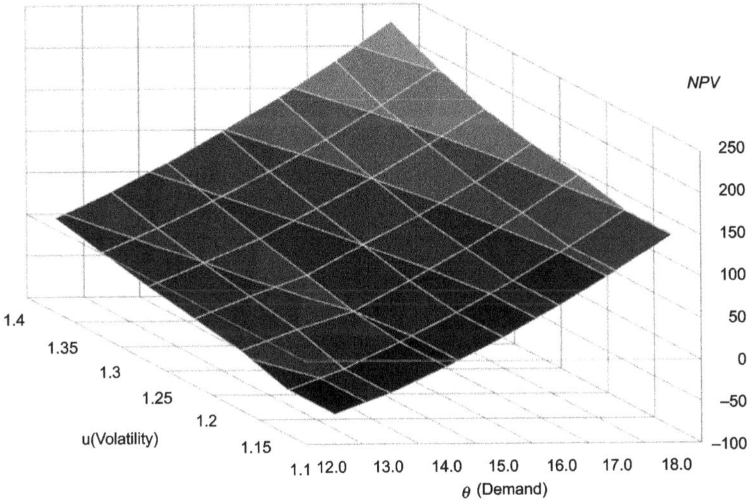

Source: Prepared by the author.

In order to build such a win-win relation, it seems useful to refer to the conditions of each NPV of a base case, a proprietary strategy, and a sharing strategy discussed here.

Conclusion

In 3-dimension models on NPV of both quantitative and pricing competitions, demand and volatility can contribute to a rise of NPV. From each type of option-game trees, it was found that the demand is specially related to the commitment, and the volatility is so to the flexibility. From both parameters, it becomes possible to select the optimal strategy to maximize NPV on game theory and real options.

It was understood that a proprietary strategy works profitably for a pioneering company on a change of cost, demand, and volatility for both competitive styles. A sharing strategy provides for a pioneering company an opportunity loss from a free-ride company by a non-investing company. However, it is possible to receive a net return from investing, depending on the conditions of demand and volatility, while ignoring a free-riding. In addition, it may be said that a sharing strategy is basically superior to a proprietary strategy from Pareto optimum, if summing up both NPV*s* of a pioneering company and a non-investing company. Therefore, it is advisable to select a sharing strategy with partnership, through a method of preliminarily preventing a free-riding, and then redistributing the total returns among the parties. For that reason, it may utilize results here as a preparation for forming guidelines.

As investment in basic research of life science is increased by government as the United States, Europe, Japan etc., there are more technology resources available for the next generation of industries. Otherwise, large pharmaceutical companies have expanded R&D resources by repeated M&A (Merger and Acquisitions), and have concentrated these on the clinical development with increasing number of test samples, being pressurize on the safety requirement by experts, complete pipeline from investors, and low cost by patients and Medicare. As a result, biotechnological start-ups become in charge of technology transfer as an applied research in a transition period of a drug-development process. There are few biotechnological start-ups possible to independently develop drugs, except for Genentech or Amgen. Therefore, in comparing the flexibility value of decision making, to overcome the death-valley and the commitment value of deterring an entry of a rival, it is necessary to further distinguish the commitment between the proprietary strategy and the sharing strategy of R&D findings. In other words, it is necessary to decide the optimal switch timing from proprietary strategy to sharing strategy to maximize NPV for the company concerned.

Basically, if volatility is high, so is the flexibility, and if demand is high, the commitment is respectively selected from the viewpoint of optimization. Furthermore, only when both parameters between demand and volatility are over a certain criterion or when a total of NPV*s* of both players, under sharing strategy is higher than under proprietary strategy, the switch is possible from a proprietary strategy to a sharing strategy. In particular, it may be said that a sharing strategy is a course to open up a possibility to the open innovation.

Here, a survival strategy of biotechnological start-ups is examined, by expanding a concept of real options on the flexibility of decision making for overcoming the death-valley into a new concept of option-game as the decision making even including the commitment. In addition to the flexibility for uncertainty and the commitment as correspondence to rivals, it is necessary to develop a strategy of prediction, evasion, control, and practical use of a risk by a modeling of risk in itself in the future. Furthermore, by referring to a study analyzing growth options of start-ups from a viewpoint of jump valuation,[16] an optional analysis in R&D can be newly tried.

Notes

*This work is supported with a Grant-in-Aid for Exploratory Research #18653031 by the Japan Society for the Promotion of Science. This paper was previously published at *Journal of Advances in Management Research*, 5 (I), 2008, 28–45 and modified for this book.

1. National Science Foundation, 'Science & Engineering Indicators 2006', *Division of Science Resources Statistics*, 2006.
2. Burrill and Company, 'Biotech 200', San Francisco, CA: Burrill Life Sciences Media Group LLC, 2007.
3. Timmons, J. A., *New Venture Creation: Entrepreneurship in the 1990s*, Third edition,Homewood, IL: Irwin, 1990.
4. Cox, J., S. Ross and M. Rubinstein, 'Option Pricing: ASimplified Approach', *Journal of Financial Economics,* 7(3), 1979, 229–64.
5. Myers, S. C., 'Finance Theory and Finance Strategy', *Interfaces,* 14(1), 1984, 126–37.
6. Mason, S. P. and R. C. Merton, 'The Role of Contingent Claims Analysis in Corporate Finance', in E. I. Altman and M. Subrahmanyam (eds), *Recent Advances in Corporate Finance, Homewood,* IL: Richard D. Irwin, 1985, 7–54.
7. Kasanen, E. and L. Trigeorgis, 'Flexibility, Synergy, and Control in Strategic Investment Planning', in R. Aggarwal (ed.), *Capital Budgeting under Uncertainty*, Englewood Cliffs, NJ: Prentice-Hall, 1993, 208–31.
8. McDonald, R., and D. Siegel, 'The Value of Waiting to Invest', *Quarterly Journal of Economics*, 101(4), 1986, 707–27.
9. Myers, S. C. and S. Majd, 'Abandonment Value and Project Life', *Advances in Futures and Options Research*, 4, 1990, 1–21.
10. Fudenberg, D. and J. Tirole, 'The Fat-Cat Effect, the Puppy-Dog Ploy, and the Lean and Hungry Look', *American Economic Review*, 88, 1984, 361–66.

11. Trigeorgis, L., 'Anticipated Competitive Entry and Early Preemtitive Investment in Deferrable Projects', *Journal of Economics and Business*, 43(2), 1991, 143–56.
12. Kester, W.C., 'Today's Options for Tomorrow's Growth', *Harvard Business Review,* 62(2), 1984, 153–60.
13. Kester, W. C., 'Turning Growth Options into Real Assets', in R. Aggarwal(ed.), *Capital Budgeting under Uncertainty*, Englewood Cliffs, NJ: Prentice-Hall, 1993, 187–207.
14. Smit, H. T. J., and L. A. Ankum, 'A Real Options and Game-Theoretic Approach to Corporate Investment Strategy under Competition', *Financial Management* 22(3), 1993, 241–50.
15. Smit, H. T. J. and L. Trigeorgis, 'Strategic Investment: Real Options and Game', Princeton, NJ.:Princeton University Press, 2004.
16. Willner, R., 'Valuing Start-up Venture Growth Options', in L. Trigeorgis, (ed.), *Real Options in Capital Investment: Models, Strategies, and Applications*, Westport, Conn.: Praeger, 1995, 221–39.

References

Accounting Statistics Related with Biopharmaceuticals Companies. *On-line Wall Street Journal,* 2006, http://markets.wsj.com/us (accessed 31 December 2006).

4

Commercial Spinoffs Generated by Italian Universities: An Empirical Analysis

Maria V. Ciasullo and Giulia Monetta

The aim of this article is to analyze the role that universities can play in contributing to the development of spinoffs. The characterizing element is the link spinoffs have with universities. The link is not limited only to the incubation phases, but continues through the whole life cycle of spinoffs, facilitating their development and survival. This link was analyzed in spinoffs generated by Italian universities and financed by the Italian law on industrial research.

The article provides conceptual arguments, and both primary and secondary empirical evidence. The results obtained from the overall analyses, allow us to confirm that the link with universities is a necessary, but not sufficient condition, for the consolidation of spinoffs. Finally, the article presents limits and possible implications for future research.

Overview

Since the 1980s, in most European countries and in the United States of America (USA), universities have focused on technology transfer to boost co-operation between research and industry. Many universities are now in the active technology-transfer phase, having undertaken agreements and alliances with large industrial groups for the development and commercialization of the results of their scientific research, especially in the USA and northern Europe.

However, only recently have the Italian universities shown interest in economically exploiting the commercial value of scientific research. On the other hand, in some sectors and geographical areas, this process has led to the launch of small- and medium-sized firms, characterized by a high level of innovative skills. In other words, an ideal environment has been created for new firms based on the ability to sell and exploit the value of research results. The university spinoff, a new company created to exploit intellectual property originating in an academic institution[1] is a case in point. However, companies established by current or former members of a university, not commercializing intellectual property, originating in academic institutions, obviously, are not included in our definition of a spinoff. Research-spinoff firms play a specific role in the knowledge market, as compared to spinoffs that are strictly industrial in nature (i.e., the offshoots of other firms); they provide a mechanism of knowledge transfer stemming from pre-competitive research to industrial applications. In short, two main factors distinguish university spinoffs: the status of individuals involved in the new business venture process (researchers or students); and the kind of knowledge transferred from the university to the new venture (codified or tacit) characterizing the university spinoff activities (product or services-oriented).[2]

Research Purpose and Objectives

The article begins from the premise that by means of applied research and the development of technology, Italian universities play a fundamental role in the processes of industrial restructuring, and social and economic development. Universities can support technological innovation and exploit the value of academic research in many ways. This article aims to examine the decisive role of universities, with their research facilities, in influencing the creation of spinoffs. A characterizing element of university spinoffs is, therefore, the link the newly formed companies must have with universities. This is fundamental, not just during the incubation phase, but also during the start-up phase, explicated in terms of providing resources that facilitate start-up and initial development and later during the life-cycle of the spinoff, to ensure firm viability and survival.

In Europe, generally, support programmes for the creation of firms using technology outputs from academic research involve a growing number of universities. In Italy in particular, up to the year 2000, no specific policy to boost technology transfer in its various forms was in place. Consequently, few strategies for stimulating the value and commercialization of research results were devised. However, during the past decade and despite the number of obstacles experienced during the start-up phase, not to mention problems of growth, the number of commercial spinoffs has increased remarkably. Undoubtedly, the new national research system introduced into the Italian legal system (Legislative Decree no. 297/1999 and Ministerial Decree no. 593/2000) has played a significant role in this slow but promising growth trend.

The article attempts to describe and explain the formation of university spinoffs and account for their role in the commercialization of university technology and in the creation of wealth and added value in Italy, by means of a detailed analysis of the spinoffs generated by Italian universities, and financed by the Italian laws on industrial research during the time frame 2001 to 2009.

In particular, the objectives are:

1. to identify resources (organizational, human, technological, physical and financial) that are indispensable for transforming a business idea into an effective business project;
2. to highlight the major obstacles that entrepreneurs encounter during the various phases of the process, i.e., from the conception of the idea (incubation) through the start-up phase, to the consolidation of the business in the marketplace;
3. to identify the contribution (forms of support, interaction, etc.) by which universities have paved the way ahead for business development.

Structure of the Article

From an overview of the literature, the study proceeds with a brief description of the Italian regulations supporting industrial research, illustrating the modalities and procedures set out in the Legislative Decree (n°. 297/1999 and Ministerial Decree n° 593/2000), for financing of university spinoffs.

Through studies of spinoffs financed in compliance with the said laws, the article attempts to measure and quantify company performance, examining the various modalities of interaction between universities and the companies involved. Based on research objectives, hypotheses were then posited.

Finally, on the basis of the data gathered (using both primary and secondary sources) findings are presented and the implications for future research underlined.

Literature Review

From an overview of the literature, four main elements, distinguishing university spinoffs emerged and a theoretical framework was identified:

1. A great many Italian and international academics have focused on the modalities of transferring university knowledge to spinoffs and have investigated the same. Some studies and research highlight the benefits that university spinoffs derive from academic collaboration inasmuch as they are distinguished by a high tendency towards innovation, and are set up in the same location in which scientific knowledge, developed inside research laboratories, is then transferred into know-how for the formation of highly competitive product offers.[3] At the same time, spinoffs help universities in three fundamental ways — they provide financial support for university research;[4] help to attract and retain a productive science and engineering faculty[5] and contribute to education and facilitate training of students.[6]
2. Various studies on innovation have pointed out that university spinoffs are classified as being new technology-based firms, i.e., younger and more innovative firms that develop and use new technology.[7] In particular, a lot of empirical evidence suggests that university spinoffs tend to be founded to exploit the following types of technology innovation — radical, because a single product extension that is merely only incremental does not justify the creation of a company;[8] tacit knowledge, because it is difficult for anyone other than the inventors (such as individuals in

established companies) to develop the knowledge into a commercial technology;[9] in the early stages of invention, when a university technology is at a very early stage of development, it cannot easily be licensed to established firms but tends to lead to the creation of spinoffs; high value for the customers, when a technology offers significant improvement in customer value, the creation of a spinoff is justified as it attracts interested shoppers and motivates inventors;[10] strong intellectual property protection, spinoffs are more likely to be founded when a large portfolio of broad scope patent protects an invention.[11]

Besides, university spinoffs are very high-performing companies and their survival rate is also extremely high. Empirical evidence in the USA,[12] in France,[13] in Sweden,[14] and in northern Ireland,[15] suggest that university spinoffs are more profitable, on an average, than the typical advanced technology start-up.[16]

3. University spinoffs encourage local economic development and they contribute significant economic and social value to the territory in which they operate, providing as they do, direct and indirect contributions to the generation of knowledge, and to the diffusion of technology. They create productivity, new jobs, and well-being. In particular, it is possible to identify various ways in which spinoffs encourage local economic activity.

Firstly, they generate significant economic value by producing a product offer capable of satisfying customer wants and needs. In this respect, university spinoffs can be seen as useful entities in finding solutions to unsatisfied customer demand. While there is no research estimating the indirect effects of the economic impact of university spinoffs in a particular area, there is some case study evidence regarding specific countries.[17]

University spinoffs are also important because as mentioned previously, they create jobs. Empirical evidence from countries, such as the United States, the United Kingdom, Belgium, Sweden, etc., highlight the greater capacity for job creation of university spinoffs.[18] Moreover, the jobs created by spinoffs are generally very knowledge-intensive and involve highly educated human resources.[19]

Lastly, university spinoffs with their domino effect have the impetus to motivate the potential innovation and entrepreneurial skills of other operators in the area.[20] They enhance economic growth by transforming university technology into business opportunities. In fact, spinoffs frequently serve as catalysts for the formation of geographic clusters of new firms in particular technologies.[21]

4. University spinoffs tend to have difficulty in obtaining financial resources. This appears more evident in the analysis high-tech firms make concerning the choice and use of various sources of capital.[22] Empirical evidence furthermore, highlights that university spinoffs need to engage in both technical and market development, before they can sell new technology products or services. In many cases, the entrepreneur acquires capital from external sources, including business angels, venture capitalists and government agencies. Nevertheless, it would appear that private sector investors are often not interested in university spinoffs at their earliest stages of development because the risks facing these ventures are quite high.[23] Outcomes suggest that entrepreneurs often use public funds to develop their new ventures, both to avoid giving up large grants, and in order to obtain adequate financing during the early high-risk stages of ventures.[24]

Brief Analysis of the Italian Norms Supporting Industrial Research

At a global level, technological research and development is taking on a primary role in the process of industrial restructuring and economic and social development. In this sense, policy makers increasingly attribute greater importance both to policies and programs that support and enhance industrial research, and to specific financial operations aimed at increasing the use of the results of research activities in an economic and social context.[25] In the context of the Lisbon Framework Program, the European Commission is increasingly committed to supporting research and rewarding projects that are the result of synergic collaboration between universities, businesses, and governments.

In Italy, only in recent years has legislation gradually come into alignment with the European Commission policy directives. In particular, the government-proposed programmes supporting industrial research now include spinoffs as a useful tool not only for the creation of new jobs,[26] but also, generally speaking, for the creation of economic and social value, and to reward firms, even small firms, that are capable of presenting product and process innovation. National literature regarding aspects of policy is also somewhat behind the times[27] compared to that found in the field of international and European studies.[28]

However, despite the regulatory efforts that have been made, empirical evidence reveals a limited level of collaboration between universities and businesses, well below that found in other European countries.[29] This, furthermore, has prevented universities from playing an active role in processes of economic and social development, and has impeded the emergence and development of high tech firms.[30] It is well known that Italy is not among the top countries for technological innovation. From a structural point of view, the country's specialization in medium and low-tech sectors, and the extremely high number of small firms constitute an objective limit that slows growth, and tends to hinder the competitive evolution of the country. Consequently, external inducements, characterized by incisive action aimed at spreading a culture of innovation, over increasingly larger areas are sorely needed.

The present regulation, in favour of industrial research (Legislative Decree no. 297/1999), is financed by the Facilitating Industrial Research Fund (FAR), through the Ministry of Education, University and Research (MIUR), and includes particular funding for joint projects presented directly by firms and the world of research (i.e. universities, public or private research centres, etc.). In particular, Decree no. 593/2000, defines the procedural modalities for the concession of funding provided by the MIUR.[31] Article 11 regulates university spinoffs, considering them a new opportunity to support the emergence and development of new firms, and favours universities playing an active role in contributing to the country's social and economic progress.

The academic staff (lecturers and researchers), doctoral students, and research fellows involved in the spinoff process, are free to act autonomously or in collaboration with their university, public or private research institutions, industrial organizations, insurance companies, mutual funds and other financial intermediaries. In the event, the applicants (university professors/researchers), subject to obtaining tenure, can be given paid leave to transfer provisionally to the spinoff in question. The transfer period cannot exceed four years, and can be renewed only once. The academics however, continue their working relationship with the university, and in addition to their salary receive a supplementary payment from the spinoff. On the other hand, the university receives contributions from the MIUR to appoint extra staff to replace those on leave. To set the process in motion, candidates have to send a relevant application to the MIUR which includes a detailed description of the research project and information regarding the market of reference, as well as the new firm's technical and financial development plan. Subsequently, a Commission of Experts appointed by the MIUR approves the project and awards the new firms with a 'contribution to expenses', a non-repayable grant for the purchase or renting of premises, or tools and equipment; as well as for operating costs for personnel, consultancy fees (market and feasibility studies), or general expenses. The activities specifically considered include, industrial research (up to a maximum of 50 percent), precompetitive development (up to a maximum of 25 percent), and diffusion of highly advanced technology. If applicants are successful, the MIUR issues an authorization for payment and endorses a contract with the applicant. Payment is disbursed on the signing of the contract, to the extent of 50 percent of the grant, on the submission of statement of costs incurred (at least twice the amount of the sum is advanced). Applicants receive a further 30 percent; on verification of completion of the project (lasting an average 24/36 months) and then the final 20 percent is disbursed.

During the implementation of project activities, university spinoffs tend to report back to the MIUR, relative to technical-economic progress. Moreover, in the event of a university spinoff deciding to interrupt a project, justification has to be given and all funds received returned to MIUR.

In the three years following the conclusion of a project, university spinoffs have to supply data and information regarding their business activity by means of annual reports. These reports provide a valid method not only of monitoring activity, but also of supporting and directing policy actions. In fact, on an annual basis, the MIUR provides the government with a report, which, subdividing spinoffs according to their size, carries out an analysis of research and development investments, focussing on particular characteristics of technological innovation, of economic sectors, and geographical areas covered in the actions undertaken, as well as evaluating their socio-economic impact.

Methodology

The empirical research was carried out in two phases characterized by different levels of detail and elaboration.[32]

Variables and Data Collection

The field of reference is made up of all the commercial spinoffs that from the year 2001 applied for a grant as per Article 11 of the Ministerial Decree 593/2000. On the date the empirical survey was carried out — December 31, 2009 — 143 requests for funding had been received. The criteria used to define the selection were:

1. exclusion of rejected requests for funding due to negative appraisal on the part of the Evaluation Committee;
2. exclusion of spinoffs generated by private and public research centres. Spinoffs originating exclusively from universities only were considered;
3. inclusion of spinoffs that were approved following positive appraisal on the part of the Evaluation Committee on the date of the empirical survey.

The sample selected were university spinoffs financed by the Italian decree on industrial research on December 31, 2009, which consisted of 68 companies.

The data sources were both primary and secondary, as multiple data ensures a more accurate and thorough analysis of the phenomenon as well as the reliability and the results of the empirical research.[33] In particular, the secondary data sources, consisting of the database provided by the MIUR, included the requests presented by each single spinoff, as well as all documentation (business plan, curriculum vitae of each proponent, and report of competence gained by participating third parties) provided to support admission for spinoff financing. In the first phase of the research, the field of reference was defined and the subsequent sample selected; and the first and second research hypotheses were subjected to the falsification process.

This formed the basis for the second phase of the field research. The cognitive goal of the second phase implied the use of primary data, gathered by means of a qualitative and quantitative questionnaire, containing mostly open-ended questions sent to academic researchers/entrepreneurs by email.

In order to submit the third hypothesis to falsification, to highlight the link between university support and the achievement of better performance, we proceeded to further define the sample. In particular, the analysis examined spinoffs that went beyond the start-up phase (three years of activity) i.e., companies funded between the period 2001–06 that were still in business on December 31, 2009. From the 68 companies within the sample, we focused on 40 spinoffs. The processing of the questionnaire data enabled us to submit the hypotheses emerging during this phase of research to the falsification process.

Research hypotheses

The hypotheses that underwent falsification were defined bearing in mind the phase of the spinoff's life-cycle at that specific stage. In particular, regarding the embryonic phase (early stage), the aim was to verify, for all firms in the sample, whether:

1. Spinoffs are knowledge-intensive, that is, capable of generating and managing resources of tangible knowledge (patents, databases, etc.);
2. Spinoffs generate not only inventions but innovation, as it is recognized that they are capable of creating economic value through new ways of meeting needs.

Limiting the analysis only to spinoffs that have overcome the early stage, the aim was to verify:

3. if and how (given resources: physical, human, financial, etc.) universities contribute to the progressive strengthening of spinoffs, and whether spinoffs are robust and stable businesses in terms of their organizational structure and market opportunities.

Findings

The following conclusions illustrate the results obtained from the field research, and discuss the descriptive aim and the process of falsification of the aforementioned hypotheses.

Descriptive Analysis of University Spinoffs

The aim of the descriptive analysis was to explain the process of the creation of spinoff companies. In particular, results highlight different factors — the trend of university spinoffs financed by industrial law (Figure 4.1); the industries in which spinoffs operate (Table 4.1); the geographic distribution and identification of the different partners involved in the spinoff creation (Table 4.2); the offer content realized by spinoffs (Figure 4.2) and the innovation observed with relation to technology and the marketplace (Table 4.3).

With regard to the introduction of the law on industrial research, Figure 4.1 underlines how it is only from the year 2005 onwards that university spinoffs begin to record an increasing trend. The phenomenon can be traced not only to aspects of a

Figure 4.1 The creation of University Spinoffs 2001–09

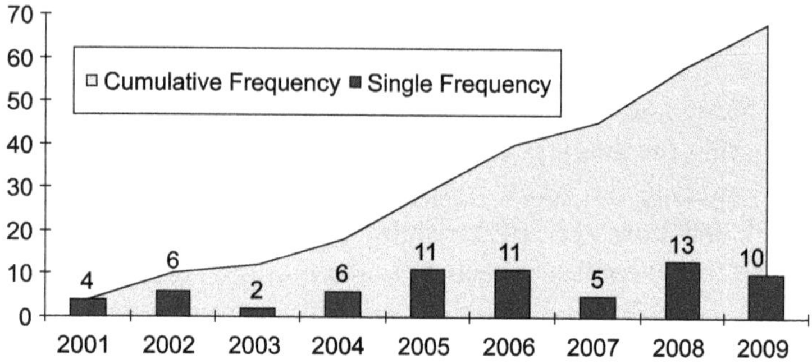

Source: All figures and tables have been prepared by the authors.

macroeconomic nature regarding, in particular, policies that have progressively helped to develop a complex system of incentives for industrial research, but perhaps even more so, to the development of an entrepreneurial university model that considers tools such as private incentives and private development in research, principal factors in sustaining scientific and economic progress. In other words, by using a 'market-oriented' approach,[34] some universities have progressively activated links and coordination mechanisms between the production of knowledge phase and that of commercial exploitation, with the aim of integrating the logic of the scientific system with the rules of the enterprise system. The said universities have supported their researchers by recognizing the business opportunities inherent in research results, and in their translation into a commercially exploitable form. The year 2007 highlighted a drastic downturn due to the world crisis which, clearly, has also generated economic-financial tension in the world of scientific research.

Moreover, data relative to the year 2009 should be considered incomplete inasmuch as from the date on which the phenomenon was noted, only spinoffs that had been approved for financing by the MIUR were considered. However, other projects were ongoing and under investigation by the MIUR.

Table 4.1 University Spinoffs by Activity Sector

NAICS Code	*Sectors of activity*	*Number Spinoffs*	*Percentage*
325	Chemical Manufacturing*	26	38
334	Computer and Electronic Product Manufacturing**	15	22
541	Professional, Scientific, and Technical Services	10	15
333	Industrial Machinery Manufacturing	7	10
311	Food Manufacturing	4	6
335	Electrical Equipment, Appliance, and Component Manufacturing	2	3
517	Telecommunications	2	3
519	Other Information Services	1	1,5
518	Data Processing, Hosting, and Related Services	1	1,5

Notes: * This industry includes firms that operate in the biotechnology field.
** This industry includes computer software firms.

To analyze university spinoff enterprises, the international activity sector classification North American Industry Classification System (NAICS) was referred. An initial consideration regarded the concentration of university spinoffs in highly innovative sectors, often accompanied by high tech solutions.

The most common industry for university spinoffs is chemical manufacturing, which includes biotechnology, pharmaceutical, and biomedical companies. One explanation for this is that the concentration of biomedical expertise lies in universities. In fact, university researchers invent a much larger percentage of patented technologies in biomedical fields than in other domains.[35] This finding is confirmed by the results of the research, where 26 of the 53 spinoffs patenting their invention are involved in chemical manufacturing.

Still, with reference to the sector mentioned in the preceding paragraph, it should be underlined that there is a significant

Table 4.2 Geographical distribution and typology of founder members of university spinoffs

Region	Individual university researcher	Several university researchers	Individual or more researchers and university departments	Individual or more researchers, industrial and/or financial partner	Individual or more researchers, University, industrial and/or financial partner	Total
Piemonte		1		1		Northern Italy 26
Liguria				3		
Lombardia	2	4		8	2	
Veneto	1	1		2		
Friuli Venezia Giulia			1			
Emilia Romagna	1	2	1	2		Central Italy 26

(Continued . . .)

(Table 4.2 Continued . . .)

Toscana	2			2		
Marche		1			2	
Abruzzo				1	1	
Lazio	1	2		5	3	
Puglia		1		2	1	
Calabria				1		Southern Italy 16
Campania				4		
Molise	1					
Sardegna			1	1	2	
Sicilia				1	1	
TOTAL	8	12	3	33	12	68

variability in the type of activities carried out by the firms. From scientific developments in genetic engineering and molecular biology, for example, there are spinoffs involved in the development of drugs, others that are focused on providing technology and technological platforms that support the industrial research and development (R&D) process, and yet others involved in the clinical experimentation of drugs.

The other activity sectors involved are computer and electronic product manufacturing, which include spinoffs involved in the production of complex data protocol and document management systems for industrial use, thus developing technologies in the field of the computerized management systems; professional, scientific and technical services, that is, firms that provide support services, also of high technological content, for the innovative activity of other firms; industrial machinery-manufacturing spinoffs concentrated on the realization of highly automated production and artificial intelligence instruments (machines and components).

Table 4.1 illustrates how most of these spinoffs concentrate on science-based sectors[36] such as, electronic, pharmaceutical, chemical and machine mechanics. Being science-driven, they are distinguished by a high level of technological change, fuelled by the spread of scientific know-how, resulting from basic and applied research activities.

It is also true that in the case of food manufacturing (a traditional sector characterized by its technological maturity), the activities carried out by spinoffs are highly innovative. In fact, thanks to developments in the field of nutritional science, and to a number of applications deriving from nanotechnology, these spinoffs produce food products which due to particular experimentation, are valid therapeutic additions for health and well-being.

In short, it would seem that the aim of spinoffs is to operate in sectors, where activity is strongly influenced by elements of overlap and circularity that exist between scientific progress, technological innovation and the marketplace.

First of all, it should be noted that all university spinoffs are established as corporations (67 firms have the legal status of limited liability companies and one has the legal status of a limited company). The geographical distribution shows a concentration wherein 76 percent of university spinoffs placed in the northern and central regions of Italy, while 24 percent are located in the southern regions. However, though there is a low concentration of spinoffs in the South, there is also a high aggregation of expertise as well as a large quantity of risk capital. In fact, 81 percent are made up of teams of at least five partners compared to 65 percent of spinoffs located in Central-Northern Italy. It should be mentioned that the quality of the team rather than the single founder, represents one of the most important factors for success, and is used as a valid indicator by institutional investors for identifying the firms with the highest chances of growth.

Looking at the taxonomy of partners, we see that in 12 percent of the spinoffs, there is the presence of an individual university researcher who is always a full-time professor. Where there is a team of academics (18 percent), as well as a full-time professor, there are also young researchers on the threshold of their careers (lecturers, Ph.D. students, etc.). In most cases, the full-time professor involved in the spinoff represents the scientific point of reference for the younger partners. The analysis of this information underlines how it is, above all, the full professors who manage valuable know-how in the groups, bring prestige to within the scientific community of reference, and possess the expertise necessary for identifying business opportunities. On the other hand, however, in contrast with the technical-scientific excellence of the team, a notable lack of managerial skills is evidenced.

In 22 percent of cases, universities participate in spinoffs as partners. These are universities committed to actively promoting the creation of spinoffs as a result of specific governance policies, aimed at strengthening the link between scientific knowledge, technological development, and application in the marketplace. In this sense, the universities provide themselves with suitable regulations in order to favour and facilitate the commercial exploitation of research results.

As concerns external partners, these are industrial organizations (private companies) and other financial intermediaries (banks). There is, therefore, an absence of insurance companies and mutual funds, although their presence is stipulated by law. Specifically,

16 percent of the partners are constituted exclusively by private companies and 50 percent by both private companies and banks. The participation of industrial organizations in the company structure ensures the presence of individuals with previous business experience and managerial skills, who generally operate in sectors dealing with similar technology or markets to those of the spinoffs. At the same time, it should be noted that industrial partners form a sort of 'cognitive fracture', inasmuch as they are equipped with innovation know-how and knowledge that cannot be compared with that of the researcher who has generated such knowledge. In other words, as they are actually not part of the research world, these partners are somewhat removed from the sources of scientific knowledge. On the other hand, however, a high percentage of spinoffs with participating banks, report positive value in terms of credibility. This at least is true for Italy, where, because of low development in the venture-capital market, traditional business initiatives meet considerable difficulties in obtaining external funding, especially in the start-up phases.

Figure 4.2 shows the offers realized by spinoffs, mapped on the basis of parameters such as, phases of the value chain, and resources necessary for the realization of the product-service.[37] Fifty percent of university spinoffs achieve final products offering solutions aimed at specific clients (35 percent), and integrated final products where the offer includes both the base product as well as a series of auxiliary services and support, aimed at fulfilling client needs in a comprehensive fashion (15 percent). The realization of offers of the type described, implies a significant commitment both in terms of resources used and phases of the value chain. In fact, production is integrated with development activities, in the sense that as well as typical research phases, it is also necessary to organize the different stages in the process regarding the physical transformation of the product, or the delivery of a service.

Spinoffs that offer know-how and services of industrial R&D make up 30 percent of the total. Such activities are intrinsically linked to the normal work of the researchers or team, and do not require significant commitment in terms of the acquisition of resources. Seven percent of spinoffs develop 'proof of concept' (prototypes and first version of products), protected by different types of intellectual property legislation. In other words, these spinoffs transform a base technology into an innovative application often requiring long periods of study, and absorbing a significant quantity of financial resources which are necessary to support the complex research. Finally, 13 percent of university spinoffs realize Original Equipment Manufacturing (OEM) components, that is, products (product parts) ready to be used by client firms. Through a process of joint design that actively involves clients, such spinoffs realize products that integrate perfectly with the clients' overall range of offers.

Figure 4.2 Distribution of university spinoffs on the basis of offer content

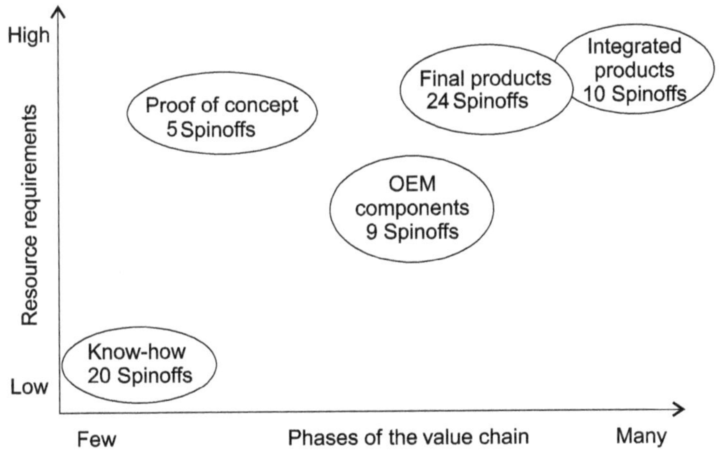

Comparing the proposed offering and the number of founder partners (Table 4.2), spinoffs making OEM components, final and integrated products, have, in most cables, a solid and well-balanced organizational structure. The involvement of professionals, both academic private, increases the credibility of these new ventures, also facilitating growth and development regards the structure, access can be had to more resources; while, concerning the dynamics, this implies enjoying increased protection relative to business aspects and managerial problems linked to the development of a new business activity. An example would be the importance for a venture capitalist or for a potential client, interacting not only with the successful academic who generated the new technology but also with an individual who is more business-minded, and has the necessary economic know-how and experience.

By cross-referencing data regarding the offer content and spinoff sector of reference (Table 4.3), the conclusions deriving from the analysis of Figure 4.2 can be confirmed. In fact, regardless of the activity sector, most spinoffs focus on realizing offers that are ready to be commercialized. Therefore, spinoffs are an effective way of transforming scientific know-how into technological solutions, processes and products that have a direct industrial application. For example, in the case of pharmaceutical spinoffs, offers are selective, personalized solutions that are effective in the cure of specific pathologies; microelectronic spinoffs realize specific automatic control systems (biosensors, control units, etc.), that are able to verify the functioning of certain industrial plants in real time; professional and technical spinoffs offer products, systems and consulting services and assistance, mainly in the field of Information and Communication Technologies (ICT), targeted at organizations whose core business is related to information technology. Moreover, the level of innovation observed with relation to technology and the marketplace, suggests, that such firms are able to fuel processes of technological subdivision, transforming generic scientific know-how into basic technology, or technology for specific application.

The analysis of the documentation shows that all firms carried out detailed market research aimed at verifying existing needs, both unsatisfied and latent. Above all, in the latter case, spinoffs were particularly effective in terms of generating innovation insofar as the business idea is concerned, and subsequently, the proposal of the offer was aimed at anticipating (or creating) the needs expressed by the marketplace. In 20 percent of the cases, market research was entrusted to a private consulting society where no such support from the university was available in this area. In all other cases, market research was carried out by specific internal university structures such as Industrial Liaison Office, Technology Transfer Office, Spinoff Point, etc. In these cases, universities offered a wide range of consulting services, aimed at evaluating effective ways of enhancing the proposed offer.

From a general perspective, it is possible to break down the activities implemented as follows:

1. technology-scouting to verify the commercial potential of the innovative product;
2. technology assessment, to verify the commercial readiness of the industrial application of the subject of innovation;
3. market assessment, for verifying the density of the potential market.

As concerns the tools of analysis — concept testing, conjoint analysis, quality function deployment and so on — increasing levels of complexity were analyzed depending on the level of know-how to be investigated.

The Support of Universities in the Perception of Researchers/Entrepreneurs

The findings from the first phase of research highlighted that the spinoffs analyzed are strongly market-oriented. They propose innovative products often aimed at satisfying specific industrial needs. This indicates that there is a high chance of product success, thus having a positive impact on firm survival. In fact, the MIUR database shows that the survival rate of spinoffs as of 31 December 2009 was 100 percent.

All spinoffs analyzed have obtained financial resources provided by the MIUR. This suggests we might expect evidence of liability of adolescence phenomena.[38] Therefore, having comfortably overcome the early stage phase (three years), thanks to the positive effect of the availability of financial start- up resources, facilitating the realization of the initial investments necessary for developing the business idea and launching it in the marketplace, in the later phases, the risk of extinction increases. Consequently, the second phase of research is focused on the one hand, on analyzing, if and how the universities of origin contributed to the

Table 4.3 Division of sectors according to Offer Content

Sector of activity	*Offer content*				
	Know-how	*Proof of concept*	*OEM components*	*End products*	*Integrated products*
Chemical Manufacturing	15	5		5	1
Computer and Electronic Product Manufacturing				8	7
Professional, Scientific, and Technical Services	5			5	
Industrial Machinery Manufacturing			7		
Food Manufacturing				4	
Electrical Equipment, Appliance, and Component Manufacturing			2		
Telecommunications				1	1
Data Processing, Hosting, and Related Services					1
Other Information Services				1	

progressive strengthening of spinoffs in the marketplace, and on the other, on verifying how far spinoffs can be considered robust and stable businesses in terms of their organizational structure and market opportunities. On the basis of these considerations, it is clear that the exploratory investigation was aimed at spinoffs that had overcome their third year of activity.

In order to submit the third hypothesis to falsification, the questionnaire was administered to 40 spinoffs. The rate of reply was 80 percent (32 spinoffs). Table 4.4 illustrates activity sectors correlated with the year in which spinoffs were founded.

Table 4.4 Activity sectors correlated with the year in which spinoffs were founded

Sector of activity	*Year*					
	2001	*2002*	*2003*	*2004*	*2005*	*2006*
Chemical Manufacturing	4	4		3	3	1
Computer and Electronic Product Manufacturing				1	2	5
Professional, Scientific, and Technical Services				1		1
Industrial Machinery Manufacturing					2	3
Food Manufacturing		2				

Regarding geographic location, of the 32 spinoffs, 28 have their headquarters in central-northern Italy and four in the south. For 15 spinoffs, following the entry of one or more industrial partners, the structure of the company is now organized differently, compared to when it was founded, and in five spinoffs, venture capitalists have also become part of the team. Moreover, it should be pointed out that universities participate in the social capital of 11 spinoffs. Another significant fact is that in regions with a higher population of spinoffs, more dynamic universities in terms of the number of firms generated, will be found.

The findings from the section of the questionnaire, aimed at investigating if and how far universities supported spinoffs, identified the main configuration of resources as well as the gaps, between the kind of support was requested by spinoffs and how much was actually given by the universities.

Entry into the marketplace of reference made it necessary for spinoffs to realize the required investments. However, as will be clarified at a later stage, requests for support were high in terms of access to physical resources such as, spaces for carrying out industrial R&D activities, use of systems and facilities, and access to laboratories and research facilities. Universities quite willingly grant the use of these resources, albeit in different ways; in fact, in 70 percent of the cases, the usage is free of charge; and 30 percent are rented to the spinoffs, however, at a price strongly undercutting current market prices.

Conversely, the factors of uncertainty and irreversibility that, in general, distinguish the realization of investments have hit spinoffs extremely badly, in particular, in the case of firms whose business model is aimed at offering products (18 spinoffs — applications, devices, etc.), where there is a significant commitment to industrialization activities. This has resulted in the decision to realize smaller investments on a preventive level for generating information and knowledge that can be used to decide whether or not to invest on a larger scale. In other words, spinoffs have put in place preliminary, minor investments in order to evaluate business opportunities more thoroughly. Moreover, the need for spinoffs to better adapt, or in some cases, optimize their proposed innovative products pursuant to client feedback, obliges spinoffs to carry out continuous developments, changes, and modifications to their technology products. It is, therefore, clear, how university facilities and laboratories form part of the investment necessary to consolidate the initiative. The motivation to improve innovative products introduced into the marketplace is the outcome of not only the feedback from the market, but also of the development of new know-how on behalf of spinoff promoters. In effect, where activities of industrialization are present, often the improved commercial exploitation of innovative products results from an increased level of know-how on the part of the spinoff researchers/entrepreneurs. Therefore, just as with physical resources, access to cognitive resources is extremely important as it underlines the need to maximize ways of penetrating open science channels. Clearly, control over know-how resources cannot be bought, but requires continuous sharing of knowledge. Above all, cases in which the researcher/entrepreneur has taken temporary leave of absence (20 spinoffs), there is a need, in short, for interaction — a constant exchange of information and communication support, aimed at guaranteeing free access to know-how developed by the scientific community.

Another kind of resource required by spinoffs is professional figures. Despite the fact they are highly motivated and are keen to create businesses, researchers/entrepreneurs have acknowledged that they do not possess adequate managerial skills, regarding operative processes and control mechanisms typical of the firm's activities. In other words, their limitations tend to become obvious, above all, in complex managerial activities; the presence, therefore, of qualified figures with managerial and organizational competencies in the team is seen as fundamental. Moreover, the researchers/entrepreneurs stated that the support offered by professional figures allows them to concentrate more on the core activities of the spinoff, which increases the productivity of R&D. Therefore, already during the foundation stage, or during the initial stages of the development of spinoffs is almost always present. As regards the involvement of businessmen as partners with adequate managerial skills. An appropriate management team facilitates the growth and viability of spinoff. In universities, researchers/entrepreneurs have privileged access to the best resources, characterized both in terms of basic education, professional growth, tendency to work well in groups, interpersonal relations, as well as a particular innate ability for analysis, problem solving, etc. They can also easily benefit from a Masters degree, and update competence in business administration, business finance, etc., as well as expand the group's know-how by involving highly qualified scientific-professional figures. Inclusion of the latter is considered be important for the absorption and use of know-how, and to incorporate scientific developments from abroad. However, such requirements are more pressing in biotechnology, pharmaceutical, and biomedical spinoffs.

In general, all spinoffs find access to financial resources complicated due to perceptions of strong risk. This often dissuades specialist investors who prefer to finance projects for the expansion of already well-established firms representing a lower level of risk, rather than invest in firms during their start-up phase. Interestingly, the analysis of responses to the questionnaire confirms that spinoffs enjoying input from financial figures (banks and venture capitalists) are part of relational networks that result from:

1. more or less direct relationships with representatives of the risk capital market, prerogative of researchers/entrepreneurs who while pursuing consultancy activities, have established contacts and relationships with the financial and economic world as well as with the scientific world (37 percent).
2. the support offered by the universities of origin whose aim is to cultivate continuous relations and contact between the academic, industrial and financial worlds so that, from the formation phase onwards, spinoffs can benefit from the contribution of industrial and financial partners, thus having a positive impact on spinoffs, both in terms of expertise and in terms of venture capital (63 percent).

At universities, spinoffs do not request financial resources as such, but hold that the participation of universities in the social capital at the funding headquarters for spinoffs the significant. When universities do participate, credibility is increased and the business project, legitimized. Moreover, simplification of the patenting process and the minimum royalties that spinoffs have to pay to the structure of origin (university, department) are also elements that have had significant impact in terms of financial savings.

It is clear how access to the resources mentioned in the preceding paragraph is characterized by a virtuous reciprocal relationship as each resource, once suitably evaluated and acquired, increases the credibility, reliability and reputation of spinoffs, thus reducing conditions of weakness. In particular, researchers/entrepreneurs consider reputation a veritable 'businesses card' for spinoffs, making as it does, a significant contribution to strengthening the business in terms of reliability and competence in the eyes of stakeholders (clients, suppliers, financial backers, new partners, local community, etc.).

The main vehicles for reputation, what we define as structural and dynamic, refer to initial assets of spinoffs at the start-up phase (structural), and the development and consolidation phases of the business initiative (dynamic). Structural vehicles can be found in the prestige of the university of origin, profile and experience of members of the business team, inclusion within the governing body of individuals enjoying renown and prestige in scientific community of reference, and individuals from the business world. Dynamic vehicles are organizational choices, and the capacity to establish and fuel positive relations as opposed to those usually found during simple market transactions. They are used to increase the perception of credibility, and therefore, of reputation. The researchers/entrepreneurs recognize the precious role played by the incubators promoted by universities, above all regarding networking services, but hold that the usefulness of the support offered is a function of the ability of management of the spinoff to fuel and enhance relations with the stakeholders. The final aim is to become a part of highly relational networks that are the result of collaborative relationships and solid partnerships.

With regard to organizational decision-making, 28 spinoffs have set up autonomous headquarters, leaving the research facility of the university of origin with the objective of locating the activity in more commercial surroundings such as, for instance, science parks, academic incubators, industrial areas, and districts and technology centres.

Quantitative analysis of the consolidation of university spinoffs

This section of the questionnaire aimed at investigating the consolidation of spinoffs in the marketplace was focused on the data gathered during the two-year period, 2008–09, relative to the following variables held to be crucial for growth processes

— employees and turnover. Being a quantitative analysis, it was thought best to exclude spinoffs founded in 2006, given that in terms of absolute value, the number of team members and turnover variables are significantly below those reported by spinoffs in business for several years. Moreover, worthy of note is that, both in terms of employees and turn over, the absolute values gathered regarding spinoffs founded in the years 2001–02 are proportionately higher than for spinoffs founded in later years.

Lastly, with reference to the size of spinoffs, the overall analyzes carried out confirm that we are dealing with quite small firms aimed at narrow client segments.

Figure 4.3 Average number of employees by sector of activity: year 2008

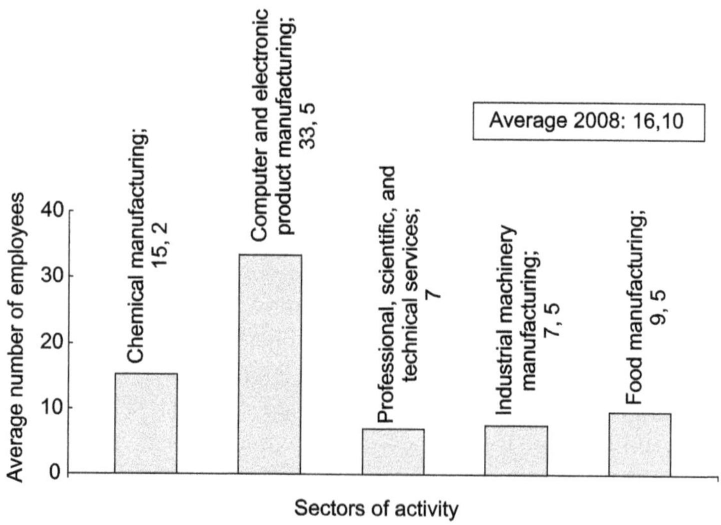

The size dimension of distribution of spinoffs (expressed in average number of employees) can be divided by activity sector. With reference to the year 2008 (Figure 4.3) larger firms are found within the computer and electronic product manufacturing sector, followed by that of chemical manufacturing. The latter sector of activity, however, has a slightly lower value than the sample average (16,10). The lowest values, which are nevertheless similar, concern professional, scientific and technical services, and industrial machinery manufacturing sectors. Food manufacturing has a higher value than the abovementioned sectors, but is still below the sample average. With reference to the year 2009 (Figure 4.4), the most important sector of activity remains that of computer and electronic product manufacturing, followed by chemical manufacturing which with an average of 18, 4 team members, is more or less in line with the sample average (18, 62). Professional, scientific and technical services show no change compared to 2008, but evidence a drop compared to the sample average. Food manufacturing and industrial machinery manufacturing show the same average rise in team members compared to 2008.

At an aggregate level, the trend for the average number of team members (Figure 4.5) shows a rise (+15, 65 percent) and is evidence of a consolidation process underway for such spinoffs.

Figure 4.4 Average number of employees by sector of activity: year 2009

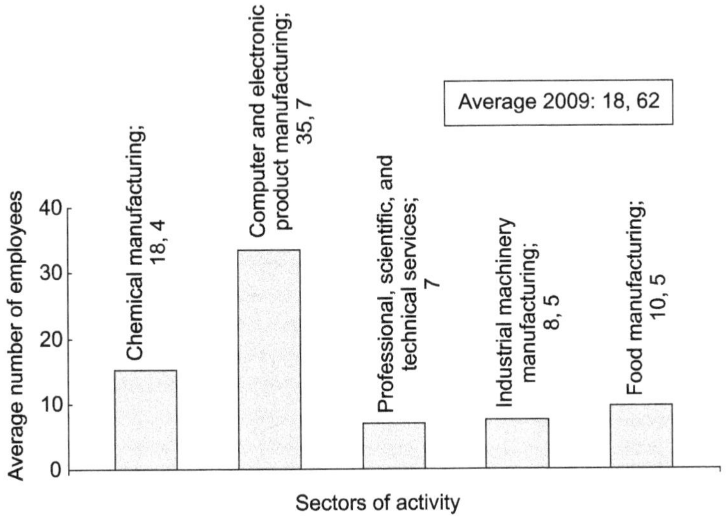

Figure 4.5 Average trend

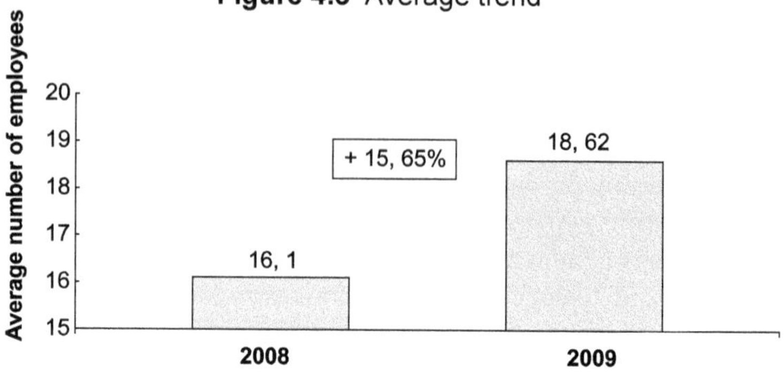

During the period measured, the structural make-up of management levels showed no change in 12 spinoffs. However, the professional figures taken on in managerial roles, vary, depending on the activity sector. For instances, chemical manufacturing, computer and electronic product manufacturing have a management team comprising five to seven individuals whereas professional, scientific and technical services, industrial machinery manufacturing, and food manufacturing on the contrary comprise two to four.

Figure 4.6 Turn over by sector of activity: year 2008

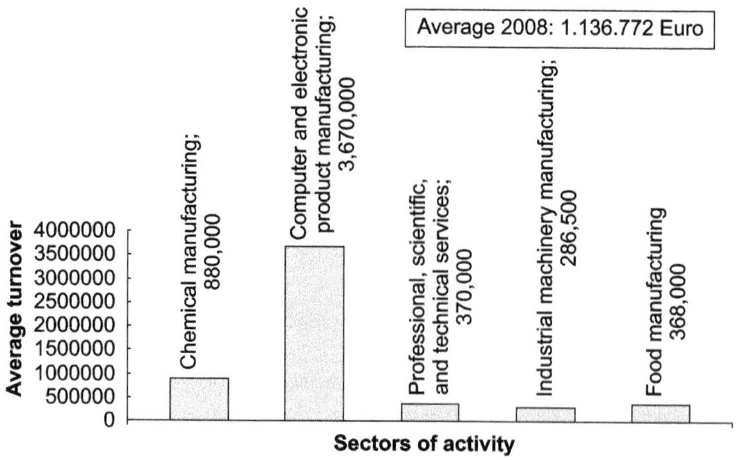

Also, the average turn over varies depending on the activity sector. With reference to the year 2008 (Figure 4.6), spinoffs with

Figure 4.7 Turn over by sector of activity: year 2009

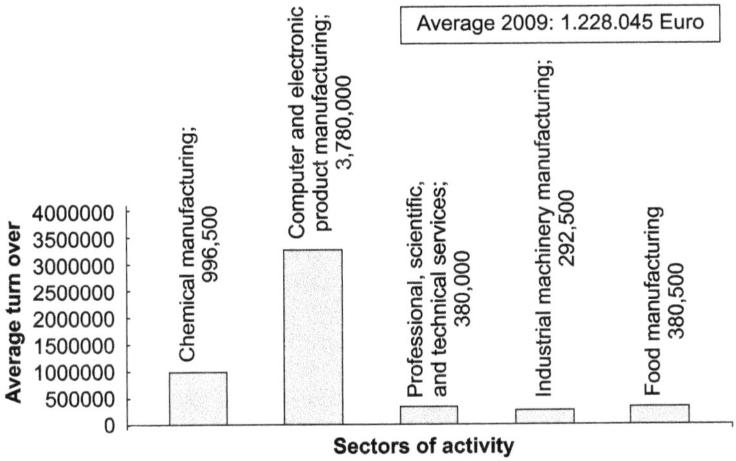

Figure 4.8 Average trend

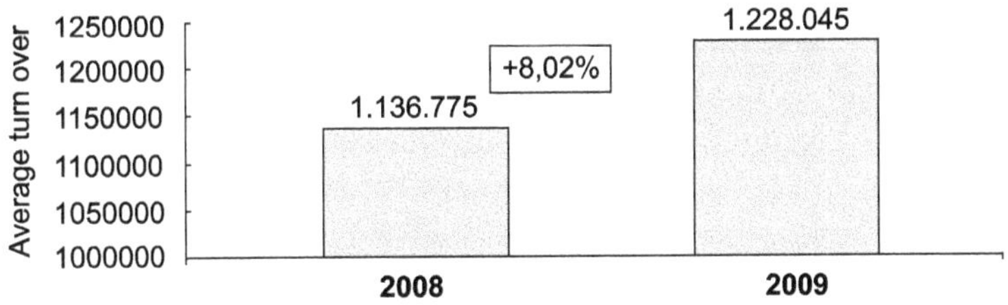

a turn over above average operate in the computer and electronic product manufacturing sector. Chemical manufacturing is the second (largest) sector in terms of average turn over; however, it is positioned below the average value of the sample. Spinoffs operating in the professional, scientific and technical services, and food manufacturing sectors have very similar values. The lowest value is regarding industrial machinery manufacturing. Moreover, the gap between computer and electronic product manufacturing, and the other sectors is significant. With reference to the year 2009 (Figure 4.7), although the average turn over value for each activity sector was in growth, trends were the same as those observed in 2008. At an aggregate level, the trend of average turn over (Figure 4.8) shows a positive tendency (+8, 02) and is still growing.

Conclusion

In short, it is possible, from a partial perspective, to draw certain conclusions from the field analysis.

All spinoffs are knowledge-intensive both in terms of localization and in generating new know-how, seeing as how intellectual property that triggers the creation of spinoffs derives from a particular cognitive asset (scientific research activities). With regard to localization, the critical factors for success are embodied in the possession of know-how (prior knowledge) and in the particular experience gained by the researcher or team that has given life to the new venture. With regard to generating new know-how, this is because the cognitive asset, brought by the researcher or team, is expressed and enhanced though the gradual transfer to the overall packet of offers, all of which contain some element of uniqueness. It is, therefore, clear how technical-scientific know-how embodied by each academic researcher, is the 'incipit' that fuels the invention-innovation process virtually. In other words, the knowledge gained from science is replicated in technical/technological solutions that help meet specific (often industrial) market targets effectively. Moreover, in a chainlike process, its scientific know-how fuels the ability of the spinoff to absorb further know-how relevant for the development of innovative products from the outside. As confirmed by Fontes[39], researchers and scientists promoting spinoffs have the advantage of preserving inside knowledge, that translates into an in-depth understanding of the potential scientific-technological developments of specific know-how, underpinning economic progress. Furthermore, the data analysis underlines how the business model adopted by spinoffs envisages and predicts real opportunities in the market, consolidating business initiatives when, besides the initial supply of resources, specific relational capacities are also present. In fact, with regard to the support supplied by universities to consolidate these business initiatives, it should be noted that raised cultural awareness of the academic context, aimed at enhancing research by means of the creation of spinoffs, has obliged universities to set up specific internal structures that guarantee the supply of resources and services, in order to 'protect' these ventures. At the same time, the spinoffs can benefit, not only from particular tangible resources, but are also encouraged to become an active part of relational networks to help find new stimuli, commercial opportunities and sources of technological know-how. Nonetheless, the support supplied by universities should not be seen as 'welfare policy', as the researchers/entrepreneurs themselves have stated not only how the particular relational abilities of management contributes to the development of spinoffs, but also those of the entrepreneurial team. Relational abilities are seen as necessary both for guaranteeing the effective coordination and integration of resources, activities and different roles relating to intra-firm aspects, and for building and enhancing relations with the other socio-economic partners. The results obtained from the quantitative analyzes, despite the small sample and heterogeneous activity sectors, confirm that these are businesses enjoying growth within their sector.

Discussion and Implications for Further Research

Rather than a point of arrival, our findings represent a starting point, i.e., a discussion point for further investigation.

If the cognitive asset is the fundamental incipit that fuels the process from which spinoffs derive, consolidation, on the other hand, is the result of the capacity to establish profitable relations with both the marketplace and other relevant stakeholders. In this perspective, dynamic networking capacity not only provides access to critical resources but also facilitates the development of relationships characterized by increasing levels of trustworthiness and credibility. In line with the literature,[40] the implications for further research and study could consist in measuring the fundamental business capacities and skills, for the generation of networking dynamics. Moreover, this would enable the verifying of the contribution of spinoffs to the development of their geographical areas of reference.

These conclusions however, prompt us to highlight the limitations of the research; in particular, by submitting the questionnaire to researchers/entrepreneurs, we have been impeded in grasping factors such as relational dynamics, particularly relevant in spinoffs that are highly complex from a technical and managerial point of view. Moreover, the analysis of networking dynamics, relative to the support offered by universities, needs to be implemented by means of a cross survey administered to relevant organs within internal university structures.

Notes

1. Shane, S., *Academic entrepreneurship: University Spinoffs and Wealth Creation*, Northampton: Edward Elgar, 2004.
2. Pirnay, F., B. Surlemont, and F. Nlemvo, 'Toward a typology of university spinoffs', *Small business economics*, 21(4), 2003, 355–69.
3. Bagby, J. W., and K. Rao, 'Strategic advantages in management of university related star-ups: Issues and Challenges', The Pennsylvania State University Center for Enterprises Creation and Growth, 1990; Carlsson, B., 'University, entrepreneurship and public policy: Lessons from Abroad', in S. Shane (ed.), *Economic development through entrepreneurship, Government, University and Business Linkages*, Northampton: Edward Elgar, 2005, 198–217; Bonaccorsi, A. and A. Piccaluga, 'University-industry relationships', *R&D Management*, 24(3), 1994, 229–47; Monjon, S., and P. Waelbroek, 'Assessing spillovers from universities to firms: evidence from French firm-level data', *International Journal of Industrial Organization*, 21(9), 2003, 1255–70; Amendola, G., 'L'imprenditorialità difficile: la creazione in Italia di high tech da parte di ricercatori universitari', in F. Martinelli and G. Bartolomei (eds), *Università e Tecnopoli* (Universities and Technology Parks), Pisa: Tacchi, 1992; Antonelli, A., *Organizzare l'innovazione. Spin off da ricerca, metaorganizzazioni e ambiente relazionale(2)*, Milano: Franco Angeli, 2004; Baroncelli, A. (2001). 'Spinoff e spinoff accademici. Definiamo i concetti'(3), in A. Baroncelli (ed.), *Percorsiimprenditoriali generati dall'Università. Il fenomeno degli Spinoff accademici*(4), Bologna: Clueb, 2001, 29–49; Roberts, E. B. and D. E. Malone, 'Policies and Structures for Spinning off New Companies form Research and Development Organizations', *R&D Management,* 6(1), 1996, 17–48; Mustar, P., 'Spinoff

enterprises: How French academics create high-tech companies: conditions for success or failure', *Science and Public Policy*, 24(1), 1997, 37–43.

4. Blair, D., and D. Hitchens, *Campus Companies-UK and Ireland*, Aldershot, UK: Ashagate, 1998.
5. Powell, W. and J. Owen-Smith, 'Universities and the market for intellectual property in the life sciences', *Journal of Policy Analysis and Management*, 17(2), 1998, 253–77; Matkin, G., *Technology Transfer and the University*, New York: Macmillan, 1990.
6. Richter, M., 'University scientists as entrepreneurs', *Society*, July–August, 4, 1986, 80–83; Etzkowitz, H., 'Research groups and "quasi-firms": The Invention of the Entrepreneurial University', *Research Policy*, 32(1), 2003, 109–21; McQueen, D., and J. Wallmark (1991). 'University technical innovation: Spinoffs and patents in Goteborg, Sweden', in A. Brett, D. Gibson and R. Smilor (eds), *University Spinoff Companies*, Savage, MD: Rowman and Littlefield Publishers, 1991, 103–15.
7. Amendola, 'L'imprenditorialità difficile: la creazione in Italia di high tech da parte di ricercatori universitari'; (Entrepreneurship Difficult); Baroncelli, A., 'Spinoff e spinoff accademici. Definiamo i concetti', Fontes, M., 'The Role of Entrepreneurial Firms in the Transfer of Public Research to the Productive Sector', Gent: Entrepreneurship Research Conference, May 21–23, 1998 (Antonelli, Organizzane l' innovazion. Spinoff da riclnca, metaorganizzazioni e ambiente relazionale' (Organizing innerates).
8. Shane, S., 'Technology opportunities and new firm creation', *Management Science*, 47(2), 2001, 202–20; Tornatzky, L., L. Waugaman, L.Casson, S. Crowell, C. Sphar, and F. Wong ,'Benchmarking best practices for university-industry technology transfer: Working with start-up companies', *A Report of the Southern Technology Council*, Atlanta: Southern Technology Council, 1995.
9. Vohora, A., M. Wright and A. Lockett, *The formation of high-tech university spinouts: The role of joint ventures and venture capital investors*, Nottingham: Nottingham Business School, 2002; Jensen, R., and M. Thursby, 'Proofs and prototypes for sale: The tale of university licensing', *American Economic Review*, 91(1), 2001, 240–59.

10. Roberts, and Malone , 'Policies and Structures for Spinning off New Companies form Research and Development Organizations',
11. Teece, D., 'Profiting from technological innovation: Implications for integration, collaboration, licensing and public policy', in D. Teece (ed.), *The Competitive Challenge*, Cambridge: Ballinger, 1987, 285–305; Nelsen, L. ,'The lifeblood of biotechnology: University- industry technology transfer', in R. Ono (ed.), *The Business of Biotechnology,*Boston: Butterworth-Heinemann, 1991, 39–75.
12. Pressman, L., 'AUTM Licensing Survey: FY 2001', Northbrook: Association of University Technology Managers, 2002.
13. Mustar, 'Spinoff enterprises. How French academics create high-tech companies: conditions for success or failure'.
14. Kobus, J., 'Universities and the creation of spin-off companies', *Industry and Higher Education*, 6(3), 1992, 136–42.
15. Blair, and Hitchens, *Campus Companies-UK and Ireland.*
16. Shane, *Academic entrepreneurship: University Spinoffs and Wealth Creation.*
17. Mustar, 'Spinoff enterprises. How French academics create high-tech companies: conditions for success or failure'; McQueen, and Wallmark. 'University technical innovation: Spinoffs and patents in Goteborg, Sweden'; Goldman, M., 'Building a Mecca for high technology', *Technology Review*, 86(3), May/June, 1984, 6–8; Cohen, W., 'Taking care of business', *ASEE Prism Online*, January 1–5, 2000.
18. Blair, and Hitchens, *Campus Companies-UK and Ireland*, Aldershot, 1998; Cohen, W., 'Taking care of business', *ASEE Prism Online*, January 1–5, 2000.
19. Blair, and Hitchens, *Campus Companies-UK and Ireland.*
20. Porcheddu, D., M. Piredda and A. Usai, 'Spin-off universitari in regioni economicamente marginali: il caso della Sardegna'(5), *Economia e Diritto del Terziario (Economics and the Lan of the Tertiary)*, 2, 2005, 645–64; Benneworth, P. S. and D. R. Charles, 'Knowledge overspill from USOs as a driver of regional development: a review', in R. Oakey, W. During and S. Kauser (eds), *New technology based firms in the new millennium,* 5, London: Elsevier, 2006, 1–23.
21. Roberts, E., *Entrepreneurs in High Technology*, New York: Oxford University Press, 1991.

22. Mansfield, E., 'Academic Research Underlying Industrial Innovation', *Research Policy*, 20(1), 1991, 1–12; Mansfield, E., 'Academic Research Underlying Industrial Innovation: Sources, Characteristics and Financing', *Review of Economics and Statistics*, 77, 1995, 55–65.
22. Baroncelli, A., V. Chiesa and A. Piccaluga, Dall'accademia all' impresa. Uno studio sulle imprese spin-off della ricerca in Italia'(6) in A. Baroncelli (ed.) *Percorsi imprenditoriali generati dall'Università. Ilfenomenodegli Spinoff accademici(7),* Bologna: Clueb, 2001, 51–70.
23. Shane, *Academic entrepreneurship: University Spinoffs and Wealth Creation.*
24. Mustar, 'Spinoff enterprises. How French academics create high-tech companies: conditions for success or failure'; Feller, I., 'An historical perspective on government-university partnerships to enhance entrepreneurship and economic development', in S. Shane (ed.) *Economic development through entrepreneurship. Government, University and Business Linkages*, Northampton: Edward Elgar, 2005, 6–32.
25. Shane, S., 'Government policies to encourage economic development through entrepreneurship: the case of technology transfer', in S. Shane (ed.), *Economic development through entrepreneurship. Government, University and Business Linkages*, Northampton: Edward Elgar, 2005, 33–49.
26. Varaldo, R. and A. Piccaluga, 'Un Ponte traindustria e Università per rilanciare la ricerca'(8). (The Entreprise), 8(8), 1994, 4–14.
27. Antonelli, *Organizzare l'innovazione. Spin off da ricerca, metaorganizzazionieambiente relazionale* (Organizing Innovation), Baroncelli, A., 'Spinoff e spinoff accademici. Definiamo i concetti'; Piccaluga, A., *La valorizzazione della ricerca scientifica* (9), *Milano: Franco Angeli* (9), 2001, Grossi, G. and P. Ruggiero, *Lo spin-off accademico: attori ed ambiente nella fase di gestazione aziendale* (10), Padova: Cedam, 2008; Malerba, F., and S. Torrisi, 'La politicapubblica' (Public Policy), in F. Malerba (ed.) , *Economia dell'innovazione (Economics of Innovation*), Carocci editore: Roma, 2005.
28. Benneworth and Charles, 'Knowledge overspill from USOs as a driver of regional development: A review'; Landabaso, M., 'Clusters and Less Favoured Regions: policy options in planning and implementations, proceedings of Dissemination of Innovation:

Clusters, Regional Instituitions and Telematics', *Thessaloniki*, 16–17 May, 2002; Cooke, P., *Knowledge Economies: Clusters, learning and cooperative advantage*, London: Routledge, 2004; Degroof, J.J. and E. B. Roberts, 'Overcoming Weak Entrepreneurial Infrastructures for Academic Spinoff Ventures', *Journal of Technology Transfer*, 29(3–4), 2004; Geiger, R., *Research and Relevant Knowledge: American Research Universities Since World War II*, Oxford: Oxford University Press, 1993; Wickstead, S., *The Cambridge Phenomenon*, Thetford: Thetford Press, 1985; Martin, B. and B. Salter, 'The Relationship between Publicly-Funded Research and Economic Performance: A SPRU Review', Science Policy Research Unit, Brighton, 1996.

29. Alessandrini, P., and A. Sterlacchini, 'Ricerca, formazione e rapporti con l'industria: i problemi irrisolti dell'Università italiana'(11), *Economia e politica industriale* (Economics and Industrial Policy), 22(88), 1995, 33–61; Ferretti, M., *Lo sviluppo delle conoscenze tecnologiche nell'impresa*(12), Torino: Giappichelli, 2007.
30. Compagno, C. 'Governance e assetti organizzativi dell'Univ ersità'(13), in M. Strassoldo (ed.) *L'azienda Università - Le sfide de l cambiamento* (The company University — The Challenges of Change), Torino: Isedi, 2001, 29–51.
31. Grossi, G. and P. Ruggiero, 'Teoria e prassi degli spinoff universitari (14), *Azienda Pubblica* (Companies Publish), xix(1), 2006, 55–76.
32. Bryman, A. *Quantity and Quality in Social Research*, London: Routledge, 1988; Agnoli, M. S., *Ildisegnodellaricercasociale*(15), Roma: Caroccieditore, 2006; Lazarsfeld, P.F., *Metodologia e ricercasociologica*(16), Bologna: Il Mulino, 1967.
33. Marbach, G., *Le ricerche di mercato* (Market Research), Torino: Utet, 1996; Metallo, G. and Barile, S., *Le ricerche di mercato (Market Research)*, Torino: Giappichelli, 2002; Chisnall, P., *Le ricerche di marketing* (Market Research). Milano: McGraw-Hill, 1994.
34. Coccia, M. and S. Rolfo, 'Technology transfer analysis in the Italian National Research Council', *Technovation*, 22(5), 2002, 291–99.
35. Kenney, M., *Biotechnology: The University-Industrial Complex*, New Haven: Yale University Press, 1986.

36. Van de poel, I., 'The Transformation of Technological Regimes', *Research policy*, 32(1), 2003, 49–68; Pavitt, K., 'Sectoral Patterns of Technical change: towards a taxonomy and a theory', *Research Policy*, 13(1), 1984, 343–73.
37. Gardner, D. M., F. Johnson, L. Moonkyu, and I. Wilkinson (2000). 'A contingency approach to marketing high technology products', *European journal of marketing*, 34(9–10), 2000, 1053–77; Mohr, J., S. Sengupta, and S. Slater, *Marketing of high technology products and innovations*, Second edition, Upper Saddle River: Prentice Hall, 2005; Compagno, C., 'Quale modello di spin-off per l'università italiana?'(17), in *Ricerca, Spin-off, Incubatori: Strategie ed Opportunità per le Università Italiane (18)*, 2° Workshop Internazionale PNI Cube, Torino, 2007, 75–80.
38. Fichman, M., and D. A. Levinthal, 'Honeymoons and the liability of adolescence: a new perspective on duration dependence in social and organizational relationship', *Academy of Management Review*, 16(2), 1991, 442–68; Bruderl, J., and R. Schussler, 'Organizational mortality: the liability of newness and adolescence'. *Administrative Science Quarterly*, 35(3), 1990, 530–47.
39. Fontes, M., 'The process of transformation of scientific and technological knowledge into economic value conducted by technology spinoffs', *Technovation*, 25(5), 2005, 339–47.
40. Walter, A., M. Auer, and T. Ritter, 'The Impact of network capabilities and entrepreneurial orientation on university spinoff performance', *Journal of Business Venturing*, 21(4), 2006, 541–67; Dagnino, G. B. (2005). 'Coupling combinative and relational capabilities in interorganizational best practice transfer: an evolutionary perspective', in A. Capasso, G. B. Dagnino and A. Lanza (eds), *Strategic capabilities and knowledge transfer between organizations: new perspectives from acquisitions, networks, learning and evolution,* Cheltenham: Edward Elgar Publishing, 2005, 103–44.

5

Organizational Factors and Innovation – Analysis of Different Types of Organizations

Vishnuprasad Nagadevara

Innovation is critical to sustaining business competitiveness and in improving productivity. The pace of technological development is accelerating and market conditions are changing more rapidly today than ever before. The innovation process is influenced by the organizational culture. This article attempts to identify the important facets of organizational culture which impact innovation. Five different categories of organizations, each having a different structure and culture are considered. The data collected from the members of these organizations is analyzed using Principal Component Analysis and resultant factor scores are calculated. These factor scores are further analyzed to highlight the differences across the five categories of organizations and their impact on innovation.

Overview

Innovation is critical to sustaining business competitiveness, and improving productivity. The pace of technological development is accelerating and market conditions are changing more rapidly today than ever before. With the advancements in the information and communication technologies, global economy has become more networked, more interdependent, and everyone is more knowledgeable about what is going on in the rest of the world. By improving productivity through innovation, businesses can increase the wealth they contribute to the economy in the form of increased employment, higher incomes, and more capital investment. Needless to say, it is that wealth that sustains public spending and leads to better education, health care, social programs, and infrastructure.

Historically, researchers have categorized innovation into exploration and exploitation. Coombs[1] suggested two sides of research and development (R&D) activities; investment mode where these activities are concerned with developing technological capabilities of organizations, and harvesting mode where R&D works with other functions of the organization to exploit special services for customers. Benner and Tushman[2] and Jansen[3] have defined 'exploitative innovations' as those which involve 'improvements in existing components and architectures and build on the existing technological trajectory', whereas 'exploratory innovations' are defined as those which involve 'a shift to a different technological trajectory'. In addition to this way of classification, authors have also indicated another dimension to distinguish between these two activities. According to He and Wong[4], an explorative innovation strategy involves 'technological innovation activities targeting new product-market domains', and 'exploitative innovation strategy' deals with technological innovation activities for ameliorating existing product-market.

Some of the authors emphasize the competence approach to exploitative innovation. They look at exploitative innovation from the point of view of competence building, as well as leveraging competence in meeting the goals of the organization. Floyd and Lane[5] argued about exploitative innovation and deploying competence. The search for new technological solutions is a continuous learning process. Cavone et al.[6] focused on the human resources of the innovating entity. The knowledge base and the skill levels that are embedded in the human resources are the key to exploitative innovation. Exploration can be viewed as the innovations at the stage of initiation or development, whereas exploitation is at the implementation stage.

The initiation stage of the innovation process has an organizational structure characterized by a high degree of complexity, low formalization, and low centralization.[7] On the other hand, an organizational structure characterized by a low degree of complexity, high formalization, and higher centralization is more amicable for the implementation stage of the innovation process. Since the initiation phase is followed by the implementation phase, it may be necessary for the organizations to change their internal structures correspondingly over time, to match the required changes.[8]

It is shown that the national culture has a strong influence on innovation.[9] It is suggested that certain cultural characteristics are more supportive in the promotion of innovation. Some cultures seem to have a comparative advantage in innovation that leads them to develop new technologies, ideas, and products. Some authors have gone to the extent of suggesting a shift of R&D activities to countries where national cultures are more amicable to innovation.[10] They argue that the R&D units located in these countries are likely to outperform those located in countries with different national cultures.

The national cultures that are more conducive to exploratory innovation could be very different from those which are more supportive to exploitative innovation. Nakata and Sivakumar[11] demonstrated the possibility of some cultures being more adept than others, in one phase of the new-product development process. Naturally, such nations should become the target for locating the corresponding innovation activities. The differences in the nature of support provided by different cultures are likely because of differentials in institutions, national policies that are supportive to the type of innovation Lundvell.[12] It has been supported by research that a country like the United States of America (USA) has been advancing in the exploratory innovation, and an Asian country like Japan has been quite successful in the exploitative innovation.

Innovation is affected by a number of organizational factors. Damanpour[13] identified factors which affect the management of innovation. Patel and Pavitt[14] showed that organizations in different industry sectors have different priorities and characteristics. These characteristics and priorities, in turn, will have a different impact on innovation activities. Different stages in an organization's life cycle may affect the type of innovation.[15] According to Utterback,[16] new technology industry is more in tune with exploratory innovation whereas a matured industry may be concentrating more on exploitative innovation.

Waterman et al.[17] argued that the relationship between strategy, structure, system, style, skill, and staff, and something called super-ordinate goals are important factors influencing innovation. Within the organization, structure manifests along multiple dimensions including centralization of authority, hierarchy of influence, and degree of role specification.[18] Tushman and O'Reilly[19] believed that organizational culture is the soul of the organization's innovation. Some of the highly innovative organizations make creativity and innovation as an integral part of their vision. Martins and Terblanche[20] showed that vision and mission are strategic determinants of organizational culture which has a great influence on innovation. Hofstead,[21] in his paper, indicated that organization culture, of late, has acquired a status similar to strategy, structure and control. Read[22] identified various determinants of innovation. According to him, the most important determinant is management support for innovation, and an innovative culture. In addition, Hofstede[23] has discussed links between national culture dimensions and the management practice in detail. Deshpande et al.[24] linked culture types to innovativeness. They have defined four types in generic culture: market culture, adhocracy culture, clan culture and hierarchical culture.

A culture which supports creativity consolidates the platform for the innovation, be it management innovation or product innovation. Organizational cultures which actively promote generation of new ideas, and creating new ways of doing things are the right platforms for creating innovation. It is important that the organizational culture should create an atmosphere where the values are shared across the entire organization.[25]

While organizational culture plays a major role, the role of the individual cannot be downplayed. Jaeger[26] attempted to link the individuals to the organizational culture. He argued that every individual carries mental programmes, that are conditioned into members of given cultural groups by their common socialization and life experiences. Thus innovation is influenced by the common programmes that are intrinsic to individuals.

It is the normal human tendency to look for a solution that is adequate for the task at hand when operating under high pressure. Amabile[27] explains that under strong external pressures to complete a task, the individual is less likely to explore new pathways. Such circumstances lead to more of incremental innovations and less of break-through innovations.

It appears that creativity, innovation, and exploration are as much individualistic as they are dependent on the organizational culture and structure. In order to be able to innovate, one needs the drive from within and support from outside. Roe[28] found that openness to experience, observance, tolerance of ambiguity, independence, autonomy, self-reliance, willingness to take calculated risks, and persistence are the attributes required for the creative behaviour in the organization. In addition, factors like sensitivity to problems, fluency, flexibility, originality, and responsiveness to feelings, motivation, and freedom from the fear of failure also affect the inventive behaviour to a large extent.[29]

Another approach to innovation is the rational goal model based on system rationalism ideology.[30] Scott[31] explained the organization as a rational system which attains productivity and efficiency by goal setting and planning. Such an organizational structure, which is driven by goal setting and goal achievements, is likely to be favouring exploitative form of the innovation. These organizations are constantly focused on the markets, leading to incremental innovations. Leadership in this type of organization provides suitable direction and initiation required for extracting maximum performance out of their employees.

On the other hand, organizations that are innovative, aggressive, adaptable and entrepreneurial seem to be best fit for the exploratory innovation.[32] Creative achievements and exploratory innovations are supported by an open organizational culture.[33] Open organizational culture challenges the assumptions of the rational goal model. The culture of these organizations is defined by informal interactions, lose control systems, low hierarchy, and high employee morale. The employees have a clear understanding of the vision and mission of the organization. People who are innovative or creative often have a low tolerance of rules and routine work, but are very comfortable with ambiguity and lack of formal structure.[34]

The study by Pandey and Sharma[35] has attempted to find patterns and themes in the literature regarding the drivers that increases both sides of the organizational creativity; exploratory and exploitative. They have explored the dimensions of organizational culture, innovation management, technology management, and leadership style which are instrumental in increasing creative output.

Morris[36] states that, [t]here appears, however, a general consensus among researchers in this field that: creativity is impeded by cultures that emphasize formal rules, respect for traditional ways of doing things, and clearly demarcated roles, creativity is encouraged by climates which are playful about ideas, supportive of risk taking, challenging and tolerant of ambiguity, democratic, participative leadership styles facilitate creativity while authoritarian styles inhibit it, creativity is enhanced by organizational structures and systems that are adaptable and flexible, access to, and effectively using, resources can be a stimulus for creativity and change'. Creativity and innovation will flourish only under certain organizational cultures and can be influenced by the several variables.[37]

About fifty organizational and general economic factors have been established to influence innovative ability.[38] Abereijo et al.[39] developed a conceptual model, based on one developed earlier by Freel,[40] to determine the innovative ability of a Small and Medium Enterprise (SME). They used nine categories. A distinction was made between manageable and non-manageable determinants of innovative ability. The determinants indicate that a number of important internal and external factors contributed, in varying degrees, to the innovation capability of small firms. An earlier study by Romijn and Albaladejo[41] also showed that the innovation capabilities of firms accumulate as a result of various internal and external inputs.

The results of the study by Abereijo et al.[42] showed that a range of factors that are both internal and external to the SMEs, were found to be statistically related to the two indices used in measuring the innovative capabilities of the companies. Among the internal factors were the higher academic degree, education in science or engineering, and previous working experience of the founder/manager(s) in large corporation/multinationals or university/research institute. In addition to these internal factors, the extent of investment on R&D and training of employees was equally important.

Hoyt, Liles and Rogers[43] highlighted the effect of selected organizational factors on creativity and innovation. They have set forth two propositions—'Organizations that demonstrate a high commitment to structured group processes would be more creative in the areas of process R&D, but organizations that provide greater autonomy (less commitment to structured group processes) would be more creative and effective at developing basic research'. They considered the proposals from a number of perspectives including group dynamics, socialization, decision making, and structure. They concluded that autonomy and independence were positively related to creativity in basic scientific research, and commitment to structured task groups produced more creativity in process-related research. The decision-making process was considered and also shown to have a moderating effect on creativity. Consensus decision-making, was shown to produce decisions with a higher level of acceptance, lower levels of creativity, and, strong on practicality. Structured conflict-decision processes, however, were shown to produce more innovative decisions that were harder to implement.

They have also concluded that creativity tends to be high in the US organizations when we consider the basic science research, but weak when we look at process and manufacturing innovation. One cause of this may be a result of the individualistic nature of the US engineers. These employees have low commitment to the organization, and are motivated by a need to continuously upgrade their skills to avoid technological obsolescence. Technological upgrading creates enhanced skills which, in turn, make the US engineers more proficient in a narrower area of science, but also improve his/her marketability to other companies.

Objectives

The innovation process appears to be influenced by the organizational factors as well as the individual initiatives. Needless to say, the individual initiatives are also affected by the organizational culture and structure. While innovation could be categorized as explorative and exploitative, the organizational factors have varying influences on these. The objectives of this study are:

1. to identify the organizational factors that influence innovation in different types of organizations;
2. to quantify, ordinally, the levels of these factors in different types of organizations; and
3. to relate the influencing factors to the type of organization.

Methodology

People are an integral part of the organization. The organizational culture is influenced by the people, and people in turn are governed by the culture of the organization. People working in the organization have their own perceptions about the organization and its ability to promote innovation. It is possible that while the policies of the organization are to promote innovation, the organizational culture is perceived to be detrimental to innovation. In other words, it is the perception of the employees about the organizational culture that will drive innovation. Based on the literature review, various factors that influence innovation are identified. These factors include autonomy, centralization of authority, sharing of vision and mission of the organization, commonality in values, etc.

These are translated into a set of questions and put into a structured questionnaire. This questionnaire is administered to randomly selected employees of different organizations. For the purpose of analysis, these organizations have been categorized into four categories.

The categories are

1. a multinational technology-oriented company (referred to as 'Multinational').
2. a research organization under complete government control (referred to as 'Research').
3. an autonomous service organization owned by the government, but having its own independent board of directors (referred to as 'Board').
4. a typical government department (referred to as 'Department').

In addition, the questionnaire is also administered to a group of persons who are from different organizations where no single organization has any influencing numbers. All these persons are working as middle-level managers. This group is used to identify the individual perceptions across different organizations having different cultures and structures. This is a sample drawn from a mixed group of companies and referred to as 'Mixed'. Thus the final analysis is carried out with five categories of organizations.

The questionnaire consists of 35 items. The respondents are asked to rate their own organization on a 5-point Likert scale, based on their own perception. Some of the items in the questionnaire are positively worded and some are negatively worded. While calculating the factor scores and the averages, responses on the negatively worded items have been retained as they were.

Principal Component Analysis (PCA) was carried out to extract various factors based on the responses. The underlying basic construct for each component is identified and further analysis is carried out across the five different types of organizations based on these components. The data on the 35 items are converted into factor scores using the Component Score Coefficient Matrix. These factor scores are used for further analysis. The analysis carried out on the factor scores involved an Analysis of Variance (ANOVA) and posthoc mean tests.

Sample Profile

As mentioned earlier, a questionnaire consisting of 35 items was developed and pre-tested. This questionnaire was administered to employees of five different types of organizations mentioned earlier. The respondents were asked to rate each statement (item) in the questionnaire on a 5-point Likert scale, based on their perception about their own organization. A total of 285 responses were analyzed for this study. The profile of the respondents is presented in Table 5.1.

Majority of the respondents were male while less than 15 percent of the respondents were female. The respondents are predominantly from a technical background. About 65 percent of them hold a technical degree while another 20 percent have a science background. Only five percent have either an arts or commerce background. Interestingly, two of the respondents have only a diploma and not a degree. Almost 50 percent of the respondents have postgraduate qualification. The postgraduate degree is predominantly either an MBA or MCA. The age of the respondents is more or less evenly distributed. The modal class happens to be the age group of 41 to 45 years. All the respondents have significant work experience. Only 20 percent have less than ten years of work experience.

Results and Analysis

Initially, the responses of the entire 285 respondents are tested for reliability using Chronbach's alpha. The value of Chronbach's alpha for the entire set of variables is 0.793. Chronbach's Alpha was also calculated when each of the items is deleted, one at a time. The minimum and maximum values for Chronbach's alpha (when the item is deleted) are 0.772 and 0.809. These values indicate that the responses are fairly reliable and none of the items need to be dropped.

Factor analysis is carried out on the 35 items. All the factors with the eigen values greater than 1.0 are extracted. Thus a total of ten factors are extracted from the 35 items, using Principal Component Analysis with Equimax rotation. The Kaiser-Meyer-Olkin Measure of Sampling Adequacy was 0.865. Generally, any value above 0.6 is considered acceptable for sampling adequacy and hence, the present value of 0.865 is more than acceptable. The ten factors that are extracted along with their underlying dimensions and the corresponding items are presented in Table 5.2.

Table 5.1 Profile of the respondents

Characteristic	*Level of Characteristic*	*Frequency*	*Percentage*
Age in Years	25–30	42	14.74
	31–35	48	16.84
	36–40	53	18.60
	41–45	68	23.86
	46–50	38	13.33
	>50	36	12.63
	TOTAL	**285**	**100.00**
Educational Background	**Type of Degree**	**Frequency**	**Percentage**
	Arts/Commerce	15	5.26
	Diploma	2	0.70
	Science	57	20.00
	Tech	184	64.56
	Post Graduate	135	47.37
	No Response	27	9.47
	TOTAL	**420**	**147.36**
Work Experience (Years)	**Years**	**Frequency**	**Percentage**
	Less than 5	22	7.72
	6 to 10	42	14.74
	11 to 15	60	21.05
	15 to 20	84	29.47
	21 to 25	29	10.18
	More than 25	48	16.84
	TOTAL	**285**	**100.00**
Gender		**Frequency**	**Percent**
	Female	40	14.04
	Male	236	82.81
	No Response	9	3.16
	TOTAL	**285**	**100.00**

Source: All tables are generated from primary data collected from respondents through a structured questionnaire.

Table 5.2 List of factors and the associated items

Factor	*Item*
Factor 1: Reward & Tolerance	Honest mistakes are recognized as an indication of courage and initiative
	People who make mistakes are encouraged to share them so that others can learn
	Effective Intrapreneurs are generally rewarded
	My organization supports employees who take risks in the interest of the organization
Factor 2: Internal Cooperation	We appoint teams from different departments to solve specific problems
	To solve a problem in one department, we would call for ideas and help from other departments
	We often build prototypes or models to rapidly test a new idea before implementing the idea
	We have a specialized unit for innovation
Factor 3: Innovation Focus	Innovation is managed centrally so that my organization can invest in projects with high potential
	Most of our innovations aimed at improving sales/revenues
	My organization can be categorized as Innovation-Driven
	Innovation is carried out systematically in my organization
	My organization uses innovation to gain competitive advantage
	My organization encourages Product Innovation

(Continued . . .)

(Table 5.2 Continued . . .)

Factor	*Item*
Factor 4: Organizational Belonging	I have a sense of belonging with my unit, but not with the organization as a whole
	Many of us are proud to be part of this organization
	When people lend a helping hand, we honour them regardless of their rank
	My organizational vision and strategies are clear to me
	People in my organization think well ahead into the future
	My organization encourages process innovation
Factor 5: Training for Creativity	Lack of proper skills is hindering innovation in my organization
	People in my organization need training in creativity and innovation
Factor 6: Centralized Control	It is very difficult to get permission to leave current assignment to join an innovation team
	Information in the organization is very closely guarded as a political resource
	The boss's ideas carry the greatest weight
Factor 7: Internal Encouragement	My organization encourages incremental innovation and not breakthrough innovation
	People in my organization always tell the truth, even when it is painful
Factor 8: Individual Initiative	Decisions in my organization are made based on understanding of customers' needs

(Continued . . .)

(Table 5.2 Continued . . .)

Factor	*Item*
	I can divert up to 50 percent of my time to explore new ideas
	Bold behaviours that lead to intrapreneurial success are punished even when the results are good
Factor 9: Adaptive Innovation	Most of our innovation is in making modifications in existing products
	Innovation in my organization is mainly directed by the top management
	We deliberately copy and adapt good ideas from outside our field
Factor 10: Self Selection	Innovation is required for the survival of my organization
	Most people who lead innovation in my organization are self-selected

As explained earlier, the scores for each of the factors is derived from the items using the Component Score Coefficient Matrix. These factor scores are summarized category-wise and presented in Table 5.3.

Table 5.3 presents the averages for each of the ten factors extracted. These averages are presented for each of the five categories namely, Multinational, Research, Board, Department and Mixed. In addition, the table presents the averages for all the categories put together. It may be noted that the factor scores for the factor 'Training for Creativity' are negative. This is because, both the items that constitute this factor are 'negatively' worded and hence, most of the component score coefficients are negative.

The Multinational category has the highest scores for the three factors namely 'Internal Cooperation', 'Innovation Focus' and 'Internal Encouragement'. These three factors involve organizational culture which was drawn from within the organization. Internal Cooperation is how the innovation issues are handled within the organization and across different

Table 5.3 Means of factor scores across different categories of organizations

Factor	*Category*					*All*
	Multinational	*Research*	*Board*	*Department*	*Mixed*	
Reward & Tolerance	3.58	3.91	3.13	2.60	3.75	3.47
Internal Cooperation	3.62	3.08	3.22	2.96	3.56	3.35
Innovation Focus	3.96	3.49	3.70	3.26	3.73	3.66
Organizational Belonging	0.24	1.31	0.98	0.07	0.44	0.61
Training for Creativity	–3.22	–3.72	–3.77	–3.23	–2.66	–3.20
Centralized Control	4.34	4.76	4.71	5.44	4.86	4.82
Internal Encouragement	3.17	2.77	2.35	2.37	2.71	2.66
Individual Initiative	1.30	0.69	0.97	0.54	1.32	1.05
Adaptive Innovation	4.32	4.60	3.96	4.53	4.18	4.25
Self Selection	5.26	5.46	5.14	5.53	5.00	5.20

departments. It also involves the way problems are handled across departments and learning that takes place. Similarly, 'Innovation Focus' involves how the organization deals with innovation. A high factor score on this indicates a high degree of focus on results and its impact on the organization. Finally, 'Internal Encouragement' involves how people are encouraged, but with focus on incremental innovation. Obviously, the breakthrough innovation is not of a major consequence here. The focus is on incremental innovations with clear impact on the performance of the organization. It has the lowest score with respect to the factor 'Centralized Control'. The employees of this organization enjoy significant amount of autonomy.

The 'Research Category' has the highest scores with respect to the three factors namely 'Reward and Tolerance', 'Organizational Belonging' and 'Adaptive Innovation'. The organization in question is a research-oriented organization which functions completely under government control. Even though this organization is termed as research organization, it has very little focus on fundamental research, but concentrates exclusively on applied research. The organizational goals are primarily on adaptation of research carried out by others in their field. Being a government-controlled organization with clear hierarchical structure; it is not surprising that the research focus is dictated by the top management. On the other hand, the organization has strong organizational belonging. Being a relatively small organization, there is a greater clarity of the organization's vision, mission and goals. As a research-oriented the organization, it has a different reward system as compared to other governmental organizations.

The 'Department' category is a typical government department involved in carrying out project-related work. Even though the projects are of a unique nature, there is very little encouragement for innovation in the organization. There is a high degree of 'Centralized Control'. The score for Centralized Control is the highest for this category. With a high degree of centralized control and little encouragement for innovation, the urge for innovation is driven only by self selection. There is very little scope for individual initiative. The organizational culture does not allow employees to divert their efforts into any creative activities. Being a monopoly department, it has very little customer orientation. The score on the factor 'Individual Initiative' has the lowest score with respect to this category. In addition to 'Individual Initiative', three other factors namely, 'Reward & Tolerance', 'Internal Cooperation' and 'Innovation Focus' also have the lowest scores. Being a typical government department, there is very little scope for rewarding any new initiatives. Also, there is very little tolerance towards mistakes and non-conformance. The logical consequence is the low focus on innovation.

The 'Mixed' category has the highest factor scores for 'Training for Creativity' and 'Individual Initiative'. The highest score for the factor, 'Training for Creativity' indicates that the employees do have the required training and the innovation activity is not held back due to lack of training or skill acquisition. There is considerable amount of freedom given to the individuals so that they can focus on innovation activity.

Finally, the 'Board' category has not scored highest on any of the factors. The organizational structure is such that it is a monopoly, owned by the government, but operates independently and autonomously. The activities of the Board are mainly in terms of service operations and maintenance. Even though the Board deals with only operations and maintenance, there is ample scope for innovations. Still, this organization scored the lowest on 'Training for Creativity', 'Internal Encouragement' and 'Adaptive Innovation'. The organization needs to initiate major changes in order to encourage innovation.

A one-way ANOVA was carried out to ascertain if there is a significant difference between means of the factor scores across the five organizational categories. The results of the ANOVA are presented in Table 5.4.

The F-Value of the Analysis of Variance varied from 2.31 with respect to 'Innovation Focus' to 10.31 with respect to 'Reward and Tolerance'. This is an indicator of the extent of differences between means. It can be concluded that the differences between means is the highest with respect to 'Reward and Tolerance' whereas the differences are lowest with respect to 'Innovation

Focus'. Nevertheless, all the F-values are statistically significant at 5 percent error level except 'Innovation Focus' which is significant at 10 percent level.

In general, Analysis of Variance indicates that at least two means out of the five under consideration are significantly different. A posthoc test is carried out to identify all the pairs of means that are significantly different using Least Significant Difference (LSD). The results of the test are summarized in Table 5.5. The numbers in the body of the table indicate the pair-wise differences in means. The asterisk indicates that the difference is significant at 10 percent level. For example, the first number indicated that the difference between the means of 'Research and Multinational' categories (Research – Multinational) with respect to 'Reward and Tolerance' is 0.3285 and that this difference is statistically significant.

There are significant differences across all the five categories of the organizational culture considered in this study. With respect to 'Reward and Tolerance', the two categories namely 'Department' and 'Board' are significantly different from the other three categories. Similarly, the 'Department' category is significantly different from all other categories with respect to 'Centralized Control'. As suggested by Cooke and Szumal,[44] centralization of authority has a major influence on 'Innovation'. The 'Department' has the highest amount of 'Centralized Control'.

The Research category has the highest score on 'Organizational Belonging'. This category is also highly oriented towards adaptive research or exploitative innovation. This supports the arguments by Martins and Terblanche[45] that vision and mission of the organization are strategic determinants of the organizational culture that influence innovation. The clarity of 'Research' category on the vision and mission of the organization influences its orientation towards 'Adaptive Innovation'.

With respect to 'Adaptive Innovation', the 'Multinational' category is very similar to all other categories whereas there are significant differences between the pairs namely 'Board and Research', 'Research and Mixed', and 'Board and Department'. The 'Multinational' category is significantly different from all

Table 5.4 One-way ANOVA for differences between means

ANOVA

Factor	*Source*	*Sum of Squares*	*df*	*Mean Square*	*F*	*Sig.*
Reward and Tolerance	Between Groups	50.08	4	12.52	10.31	0.00
	Within Groups	340.00	280	1.21		
	Total	390.08	284			
Internal Cooperation	Between Groups	16.65	4	4.16	3.85	0.00
	Within Groups	302.46	280	1.08		
	Total	319.11	284			
Innovation Focus	Between Groups	10.64	4	2.66	2.31	0.06
	Within Groups	321.80	280	1.15		
	Total	332.44	284			
Organizational Belonging	Between Groups	47.44	4	11.86	12.81	0.00
	Within Groups	259.24	280	0.93		
	Total	306.68	284			
Training for Creativity	Between Groups	63.53	4	15.88	16.40	0.00
	Within Groups	271.13	280	0.97		
	Total	334.67	284			

(Continued . . .)

(Table 5.4 Continued . . .)

Factor	*Source*	*Sum of Squares*	*df*	*Mean Square*	*F*	*Sig.*
Centralized Control	Between Groups	22.69	4	5.67	4.76	0.00
	Within Groups	333.97	280	1.19		
	Total	356.66	284			
Internal Encouragement	Between Groups	19.53	4	4.88	4.59	0.00
	Within Groups	297.91	280	1.06		
	Total	317.44	284			
Individual Initiative	Between Groups	24.78	4	6.19	5.56	0.00
	Within Groups	311.97	280	1.11		
	Total	336.75	284			
Adaptive Innovation	Between Groups	13.82	4	3.45	3.04	0.02
	Within Groups	317.89	280	1.14		
	Total	331.70	284			
Self Selection	Between Groups	11.19	4	2.80	2.69	0.03
	Within Groups	291.76	280	1.04		
	Total	302.95	284			

Table 5.5 Results of Posthoc testy for differences between means

Factor	*Category*	*Multinational*	*Research*	*Board*	*Department*	*Mixed*
Reward & Tolerance	Research	0.3285		.7820*	1.3063*	0.1577
	Board	–.4535*	–.7820*		.5243*	–.6242*
	Department	–.9778*	–1.3063*	–.5243*		–1.1486*
	Mixed	0.1708	–0.1577	.6242*	1.1486*	
Internal Cooperation	Research	–.5374*				
	Board	–0.4049	0.1326			
	Department	–.6651*	–0.1277	–0.2603		
	Mixed	–0.0658	.4716*	.3390*	.5993*	
Innovation Focus	Research	–0.4713		–0.2090	0.2313	–0.2392
	Board	–0.2624	0.2090		0.4403	–0.0303
	Department	–.7026*	–0.2313	–0.4403		–.4705*
	Mixed	–0.2321	0.2392	0.0303	.470562*	
Organizational Belonging	Research	1.0760*		0.3316	1.2446*	.8751*
	Board	.7444*	–0.3316		.9130*	.5435*
	Department	–0.1686	–1.2446*	–.9130*		–0.3695
	Mixed	0.2009	–.8751*	–.5435*	0.3695	
Training for Creativity	Research	–.5025*		0.0462	–.4929*	–1.0609*
	Board	–.5488*	–0.0462		–.5392*	–1.1071*
	Department	–0.0096	.4929*	.5392*		–.5679*

(Continued . . .)

(Table 5.5 Continued . . .)

Factor	*Category*	*Multinational*	*Research*	*Board*	*Department*	*Mixed*
	Mixed	.5583*	1.0609*	1.1071*	.5679*	
Centralized Control	Research	0.4178		0.0509	–.6805*	–0.1055
	Board	0.3669	–0.0509		–.7314*	–0.1564
	Department	1.0982*	.6805*	.7314*		.5749*
	Mixed	.5232*	0.1055	0.1564	–.5749*	
Internal Encouragement	Research	–0.4008		.4225*	0.4037	0.0664
	Board	–.8233*	–.4225*		–0.0189	–.3561*
	Department	–.8044*	–0.4037	0.0189		–0.3372
	Mixed	–.4672*	–0.0664	.3561*	0.3372	
Individual Initiative	Research	–.6117*		–0.2819	0.1533	–.6270*
	Board	–0.3299	0.2819		0.4353	–.3451*
	Department	–.7651*	–0.1533	–0.4353		–.7803*
	Mixed	0.0152	.6270*	.3451*	.7803*	
Adaptive Innovation	Research	0.2760		.6390*	0.0662	.4181*
	Board	–0.3631	–.6390*		–.5728*	–0.2209
	Department	0.2098	–0.0662	.5728*		0.3520
	Mixed	–0.1422	–.4181*	0.2209	–0.3520	
Self Selection	Research	0.1984		0.3178	–0.0688	.4609*
	Board	–0.1194	–0.3178		–0.3866	0.1431
	Department	0.2672	0.0688	0.3866		.5296*
	Mixed	–0.2625	–.4609*	–0.1431	–.5296*	

other categories, except ‘Research’, with respect to ‘Internal Encouragement’. On the other hand, the ‘Innovation Focus’ is the highest for the ‘Multinational’ category, but the ‘Organizational Belonging’ is very low. It is significantly lower than that of the ‘Research’ category. The ‘Organizational Belonging’ does not appear to impact explorative innovation.

Summary and Conclusion

This article attempted to identify the organizational factors that influence innovation in different types of organizations. After identifying these factors, an attempt is made to quantify them, ordinally, and to test for differences based on different organizational categories. Five different categories of organizations, each with a different organizational culture are selected. A structured and pretested questionnaire with 35 items was administered to the members of these organizations to elicit their perceptions about various facets of the organization. Ten factors were extracted from these 35 items using Principal Component Analysis.

The factor scores are calculated using Component Score Coefficient Matrix. These scores are subjected to further analysis to identify the differences across different categories of organizations. These analyzes involved ANOVA and posthoc tests.

The five organizations considered in this study have different cultures as well as different organizational structures. They also have different objectives and orientation. There is a significant difference with respect to the average factor scores for each of the ten factors that influence innovation. The culture and the objectives of the ‘Research’ category organization sets it apart from others with respect to certain factors. Similar is the case with the ‘Department’ category organization. The hierarchy and work culture within the organization allows very little scope for innovation. There is a certain amount of commonality between the ‘Multinational’ and ‘Mixed’ category organizations. This is because many of the respondents of the ‘Mixed’ category are from organizations which have very similar structure, culture, and work ethos as that of the ‘Multinational’ category.

Notes

1. Coombs, R., 'Core competencies and the strategic management of R&D', *R&D Management*, 26 (4), 1996, 345–55.
2. Benner, M. J. and M. L. Tushman, 'Exploitation, exploration, and process management: The productivity dilemma revisited', *Academy of Management Review*, 28(2), 2003, 238–56.
3. Jansen, J. J. P., 'Exploration and Exploitation in Technology Marketing: Building the Ambidextrous Organization', *International Journal of Technology Marketing*, 1(1), 2005, 5–6.
4. He, Z. L. and P. K. Wong, 'Exploration vs. exploitation: An empirical test of the ambidexterity hypothesis', *Organization Science*, 15(4), 2004, 481–94.
5. Floyd S. W. and P. J. Lane, 'Strategizing throughout the organization: managing role conflict in strategic renewal', *Academy of Management Review* 25(1), 2000, 154–77.
6. Cavone A. Vittorio Chiesa and Raffaella Manzini, 'Management style in Industrial R&D organizations', *European Journal of Innovation Management*, 3(2), 2000, 59–71.
7. Duncan, R. B., 'The ambidextrous organization: Designing dual structures for innovation' in Killman, R. H., L. R. Pondy, and D. Slevin (eds), *The Management of Organization*, New York: North Holland, 1976, 167–88.
8. Duncan, 'The ambidextrous organization: Designing dual structures for innovation'.
9. Kedia, B.L., Robent T. Killer and Scott D. Jullan, 'Dimensions of national culture and the productivity of R & D units', T*he Journal of High Technology Management Research*, 3(1), 1992, 1–18; Nakata, C. and K. Sivakumar, 'National Culture and New Product Development: An Integrative Review', *Journal of Marketing*, 60 (1), 1996, 155–65.
10. Kedia Keller and Jullan, 'Dimensions of national culture and the productivity of R & D units'.
11. Nakata, and Sivakumar, 'National Culture and New Product Development: An Integrative Review'.
12. Lundvall B.A.,'Innovation as an interactive process: from user producer interaction to the national system of innovation', in Dosi, G, C Freeman, R Nelson, G Silverberg, and L Soete (eds), *Technical change and Economic Theory*. Pinter Publishers, London, 1988, 349–69.

13. Damanpour, F., 'Organizational innovation: a metaanalysis of effects of determinants and moderators', *Academy of Management Journal*, 34(3), 1991, 555–90.
14. Patel. P and Pavitt. K, 'Large firms in the production of the world's technology: an important case of non-globalization', *Journal of International Business Studies*, 22(1), 1991, 1–21.
15. Utterback, J. M., *Mastering the Dynamics of Innovation*, Harvard Business School, Boston, 1996.
16. Utterback, *Mastering the Dynamics of Innovation.*
17. Waterman, R., *Adhocracy: The Power to Change*,Whittle Direct Books, Memphis, TN, 1990.
18. Cooke, R. A., J. L. Szumal (2000), 'Using the organizational culture inventory to understand the operating cultures of organizations', in Ashkanasy, N. M., C. P. M. Wilderom, M. F. Peterson (eds), *Organizational Culture and Climate*, Sage, Thousand Oaks, CA, 2000, 147–62.
19. Tushman, M. L. and C. A. O'Reilly, 1997, *Winning through innovation: A practical guide to leading organizational change and renewal*, Boston, MA: Harvard University Press, 1997.
20. Martins, E. C. et al., 'Building organizational culture that stimulates creativity and innovation', *European Journal of Innovation Management*, 6(1), 64–74, 2003.
21. Hofstede, G., *Cultures and Organizations: Software of the Mind*, London: McGraw-Hill UK, 1991.
22. Read, A., 'Determinants of Successful Organizational Innovation: A Review of Current Research', *Journal of Management Practice,* 3(1), 2000, 95–119.
23. Hofstede G., 'Riding the Waves: A Rejoinder', *International Journal of Intercultural Relations,* 21(2), 1997, 287–290.
24. Deshpande, R. John U. Farley and Frederick F. Webster, Jr., 'Corporate culture, customer orientation, and innovativeness', *Journal of Marketing*, 57(1), 1993, 23–37.
25. O'Reilly, C., 'Corporations, culture and commitment: motivation and social control in organizations', *California Management Review*, 31(4), 1989, 9–25.
26. Jaeger, A. M., 'Organization development and national culture: Where's the fit?', *Academy of Management Review*, II(1), 1986, 178–90.
27. Amabile, T. M., 'A Model of Creativity and Innovation in Organizations', *Research in Organizational Behavior,* 10, 1988, 123–67.

28. Roe, A. 1963, *Scientific Creativity*, New York: John Wiley, 1963.
29. Raudsepp, E. V., *How to Create New Idea*, Englewood Cliffs, New Jersey, Prentice Hall, 1983.
30. Barley, S. R. and G., Kunda, 'Design and Devotion: Surges of Rational and Normative Ideologies of Control in Managerial Discourse', *Administrative Science Quarterly*, 37(3), 1992, 383.
31. Scott, W. R., *Organizations: rational, natural and open systems*, Prentice Hall, Inglewood Cliffs, N.J., 1992.
32. Zammuto, R. F., B. D., Gifford and E. A. Goodman, 'Managerial ideologies, organization culture, and the outcomes of innovation: A competing values perspective', in N. M. Ashkanasy, C. P. M. Wilderom and Peterson M. F. (eds), *Handbook of organizational culture and climate,* Sage Publications: Thousand Oaks, CA, 2000, 261–78.
33. Shalley, C. E., L.L. Gilson and T.C. Blum, 'Matching creativity requirements and the work environment: effects on satisfaction and intentions to leave', *Academy of Management Journal,* 43, 2000, 215–23.
34. Ekvall, G., 'Organizational climate for creativity and innovation', *European Journal of Work and Organizational Psychology*, 5(1) 1996, 105–23.
35. Pandey, S. and R. R. K., Sharma, 'Organizational factors for exploration and exploitation: a conceptual review', *Global Business and Management Research: An International Journal* 1 (2), 2009, 1–18.
35. Pandey, S. and R. R. K., Sharma, 'Organizational Factors for Exploration and Exploitation', *Journal of Technology Management & Innovation*, 2009, 4(1), 2009, 48–58.
36. Morris (2003), Quoted in Pandey, S. and R. R. K., Sharma, 'Organizational Factors for Exploration and Exploitation'.
37. Martins, E. C. and F. Terblanche, 'Building organizational culture that stimulates creativity and innovation'.
38. Van Dijk B, R. Den Hertog, B. Menkveld, R. Thurik , 'Some new evidence on the determinants of large and small-firm innovation', *Small Business Economics*, 9(4), 1997, 335–43; St-Pierre, J. and Mathieu, C., *Innovation in Canadian SMEs: The process, characteristics of firms and their environment*, International Council for Small Business, 48th World Conference, Belfast, 2003; Romijn, Henry, 1999. *Acquisition of technological capability in small firms in developing countries,* London: Macmillan, 1999.

39. Abereijo, I. O., M. O. Ilori, K. A. Taiwo and S. A. Adegbite, 2007 'Assessment of the capabilities for innovation by small and medium industry in Nigeria', *African Journal of Business Management*, 1(8), 209–17, November 2007.
40. Freel M. S., 'Strategy and structure in innovative manufacturing SMEs: the case of an English region', *Small Business Economics*, 15(1), 2000, 27–45.
41. Romijn, Henry, Albaladejo Manuel, 'Determinants of innovation capability in small UK firms: an empirical analysis',Queen Elizabeth Working Paper Series No. 40, 2004, http://www3.qeh.ox.ac.uk/pdf/qehwp/qehwps40.pdf (accessed 2 February 2010).
42. Abereijo, I. O., M. O. Ilori, K. A. Taiwo and S. A. Adegbite, 2007 'Assessment of the capabilities for innovation by small and medium industry in Nigeria', *African Journal of Business Management,* 1(8), 209–17, November 2007.
43. Hoyt, J., D. Liles and Rogers K. Jamie, 'The Influence Of Organizational Factors On Innovation And Creativity In U.S. And Japanese High Technology Firms, 1997, http://ieeexplore.ieee.org/xpl/tocresult.jsp?isnumber=13450&isYear=1997&count=500&page=3&ResultStart=75 (accessed 2 February 2010); Hoyt, J., D. H. Liles, and K. J. Rogers, 'A comparison of group processes in U.S. and Japanese high technology firms'. In *Proceedings of the 1997 Portland International Conference on Management of Engineering and Technology,* Portland, OR, July 27–31, 1997.
44. Cooke, and Szumal, 2000, 'Using the organizational culture inventory to understand the operating cultures of organizations'.
45. Martins, E. C. and F. Terblanche, 2003, 'Building organizational culture that stimulates creativity and innovation'.

6

Networks of Small Producers for Technological Innovation: Some Models

Pankaj Chandra

Small producers face a variety of challenges — some related to markets and others related to capabilities. Inability to develop technological capabilities has often restricted small firms from growing large. In this paper, we present learning from three global networks, i.e., Technology Advanced Metropolitan Area (TAMA) in Japan, Wenzhou in China, and Rajkot in India that have adopted a variety of mechanisms of coordination between small producers, and has led to both capability enhancement and demand enhancement. We argue that the capability enhancement effects, play as significant a role as demand enhancement effects, in the growth of small firms. Coordination that allows firms to improve their capabilities enhances both productivity as well as innovative capabilities, to develop new products and processes. The paper, with the help of these three case studies, presents a generic model for small and medium enterprises (SME) development, that is based on acquiring distinctive capabilities and linkages with other small producers or other members of the supply chain. We propose distinctive determinants of a collaborative model for engaging SMEs in technological innovation over a period of time. These are: Focus of the Firm, Interactive Producers, Processing and Product Manufacturing, Innovation Investment, Markets, Market Makers (and market making processes), and Regulatory Support.

Introduction

Small producers, especially in emerging economies, face a challenge of becoming part of a capability-driven economy. While large firms internalize many of the costs associated with developing capabilities to meet the requirements of the market, small producers do not possess the scale needed to drive such activities internally. In addition, they face coordination failures due to a variety of reasons—inability to build capabilities that may be required to develop market-driven products, inability to develop channels of distribution, unable to become part of existing supply chains, unable to procure technology at discounted prices, unable to develop new processes to overcome cost barriers in the market, etc. Most simply do not have access to complementary resources needed to run their business effectively. Scarcity of working capital also means that many investments required to enhance the product quality, lead times, and productivity is not available to them easily. Moreover, the overwhelming power of channel members vis-à-vis SMEs, forces these producers to sell at razor-thin margins. This may get exaggerated when the market for the products produced by SMEs is not located in the vicinity of production units.

Small producers, especially those that might be categorized as tiny or even home workers, struggle to become part of market-driven production and distribution networks. As a result, opportunities to learn about new products required by the market, efficient processes, and effective practices are limited. Their small size, often, limits the requirements of key raw materials, hence their inability to negotiate a good price or favourable delivery schedules. As a result, their supply chain becomes extremely fragmented both in terms of supply of intermediates as well as the demand from the market. This one characteristic, i.e., requirement to develop the supplier base, simultaneously, while developing the customer base, leads to poor conversion of opportunities in emerging markets.[1] Overcoming these challenges requires understanding the logistics of supply as well as the process of accessing capabilities.

In this paper, we explore alternative models of organizing small producers to facilitate improvements in productivity and innovation on product, process and practice-related domains. These models illustrate formation and operation of various kinds of networks that help reduce the market failures that small producers face, and point towards mechanisms used to coordinate activities of a large number of small producers.

Challenges Facing Small Producers

One of the unique features of modern manufacturing in India has been the dilemma of strategic positioning of SMEs. SMEs desire to become ancillary producers of large plants whereby they can get stable demand (which implies a homogeneous product mix), but where the economics of production favours a large supplier. On the other hand, the role of SME as an innovative job shop (where the SMEs have both cost and capabilities related advantage) is something that small producers shy away from. This issue of strategic positioning is at the heart of many challenges that small firms face—whether to focus on customization, or be tied to a single customer for a guaranteed volume. Often, small producers start by developing a new/innovative product or a process, but struggle to develop a lead second or a third equally innovative product/process. In the industrial environment, one finds that many of these small producers are technically sophisticated and highly capable of innovating but constrained for resources. As can be seen from Table 6.1, the kind of manufacturing environment that supports this capability is a job shop and not an assembly line or a continuous process. In other words, SMEs have a natural cost and technological advantage (which they normally do not exploit) to provide customized service, as opposed to large producers (i.e., scope advantage as opposed to scale). Because of their inability to overcome many of the market and capability failures, many desire to become ancillaries of large units (i.e., batch process units without the advantage of scale), rather than becoming producers of variety (i.e., job shops). It becomes nigh impossible for them to have costs lower than scale producers—a highly price-sensitive

Table 6.1 The Product-Process Matrix

Nature of Production Process (examples)	*Product Characteristic*	*Process Characteristic*
Job Shop (machine shop, garage, tailor etc.)	Small volume, large variety	Process-based layouts, general purpose machines, flexible, requires innovative skills, non-standard tasks, economies of scope
Batch Process (apparel production, paint shop, press shop etc.)	Mid-size volume, mid-size to low variety	Process and product-based layout, special and general purpose machines
Assembly Line (car assembly, electronic goods assembly etc.)	Large volume, very low variety, standard product, discrete parts	Product-based layout, special purpose machines at each assembly centre, high throughput, requires process discipline, repetitive tasks, economies of scale
Continuous Process (paper, fertilizer, chemicals etc.)	Large volumes, very low variety, standard product	Product-based layout, high process control, dedicated pipelines or transport systems, high throughput, process control, economies of scale

Source: Adapted from Hayes and Wheelwright, 1979

domestic market penalizes them in the medium and long run, and shifts the balance of channel power away from them.

The above highlights the need faced by SMEs to identify areas of innovation that have market value. Those that are able to either secure support from potential customers, or are able to pursue opportunities with sustained development, or are able to ride the wave with extraordinary process capabilities, turn successful. A related pathology for SMEs is their inability to access markets, and the absence of credible market makers. While there have been agents in various sectors of industry, especially the textiles, they have not been able to span the entire supply chain or match innovative capabilities with customer requirements; the exception, however, are the 'impannatore' or the market-makers of Italy who source specific weaving, spinning or garmenting capabilities of small producers in the Prato region of Florence and Pistoia provinces to match requirements of different orders from all over the world.[2] Many of these agents lost credibility over the years, both due to their inability to deploy new practices or technology in their business to make the interface efficient for their customers, as well as due to their non-transparent business practices.

There does not exist a strong eco-system of process and product manufacturing for SMEs in India. Clusters, though developed in several parts of the country, do not have firms that cover the entire technology supply chain. These are firms of similar type who come together to share some common facilities, e.g., access to common affluent-treatment plant in a cluster of chemical plants. Industrial estates have simply become geographical agglomerates of non-interacting firms. The network of capabilities continues to remain incomplete within the cluster. It is often difficult to locate producers who will provide distinctive process support. For example, it is believed that the range of capabilities that exist in the famous Ota ward of Tokyo is so diverse and sophisticated that if one drops any blueprint, there is surely going to be some SME that will have process capabilities to develop it. Such an eco-system is largely missing in India.

Regulatory support for SMEs involved in innovative activity has not taken into account some peculiar problems faced by them. Innovation funding is still scarce especially at the seed and prototyping stages. Availability of high quality material in very small volumes or versatile job-shop-like units that will process a variety or jobs, or materials, or advanced testing facilities is quite scarce. The general research and development (R&D) orientation of these small firms, which should be their natural forte, is weak, at best.

The innovator-entrepreneur plays the role of a super-manager—managing the internal processes, markets and channels, finances and accounting, as well as getting the requisite permissions from regulatory authorities etc. — a role that leaves the innovator little time and resource to develop their technological and managerial innovations any further. Many of these challenges drive the enterprise towards survival rather than growth-driven strategies, and these disabling efforts take these firms away from becoming technological leaders in the industry — a goal with which many enterprises had been started.

The task before managers and policy makers is to design structures and strategies, aimed at removing hurdles that distract producers from innovating and improving their productivity. In what follows, we present three innovative models of cooperation that have tried to overcome many of the challenges discussed above. These examples are:

1. the TAMA Network of Japan
2. the Wenzhou Industrial Development Model in China
3. the Engineering Network of Rajkot in India

Each of these cases presents an interesting perspective on organizing small producers in some formal or informal network successfully. Next, we discuss key elements of these models and highlight the common characteristics.

The TAMA Network of Japan[3]

TAMA stands for Technology Advanced Metropolitan Area—an area in Japan that has produced a large number of innovative products and processes, through linkages between a large number of small and a few large producers. The TAMA region, an inland industrial area, covers an area of approximately 3000 sq km, spans 74 municipalities, and is home to over 10 million people, of which four million work in the TAMA network firms. Geographically, TAMA region today comprises the southwestern region of the Saitama prefecture (including Kawagoe, Sayama, Tokorozawa, Iruma, and Hanno), parts of Tokyo (includes cities like Hachioji and excludes 23 wards of Tokyo), and the central part of Kanagawa prefecture (Shonan area including Sagamihara and Fujisawa)—all comprising the adjoining prefectures of the Greater Kanto Region. Interestingly, two key highways, the Ken-O-Do Expressway and Route 16 run parallel, connect the entire length of TAMA region. In 1998, the value of goods shipped from TAMA region was USD 214 billion. As a comparison, the total value of all firms in Japan was USD 1083.883 billion while TAMA's contribution to it was USD 81.862 billion. The same figure for Holland, for example, was USD 63.89 billion. The total value-add per worker were USD 116,447 for TAMA (as compared to USD 110,1184 for all of Japan, USD 94,970 for United States of America [USA] and USD 58,260 for Germany). In the same year, TAMA had twice the shipment value of the Silicon Valley.[4] Let's now look at how such a successful performance emerged from small firms.

TAMA's roots are in the textile industry of Hachioji city in the Tokyo prefecture. By 1930, with war preparations underway, these firms joined the production of arms and related products—airplane parts, measuring instruments, and armaments—and they also moved away from Central Tokyo and the Keihin Bay Area to what is TAMA region now. It can be hypothesized that the high conformance quality of Japanese manufacturing, stems from the stringent requirements of and experience in the measuring instruments and the armament industries. TAMA region, along with the Ota Ward of Tokyo, emerged as centres of excellence in commercial engineering and processing of metals. With the end of World War II, these armament factories transformed themselves to producing non-military goods in diversified sectors like instrumentation, electronics, machine control, metal processing, machinery of all types, etc. Many workers with advanced technical skills in design, control, instrumentation, etc., started their own firms in TAMA; a transformation of engineers into entrepreneurs took place. By 1950, Japan had seen tremendous industrial growth, especially in the Tokyo bay area. The government passed laws to reduce congestion and pollution in Tokyo, thereby forcing many firms here to shift base to the neighbouring TAMA region. Subsequently, large firms started to set up their R&D centres in TAMA, while moving their mass production units to other regional locations, including outside Japan in 1980s and 1990s. Many high-skilled employees of large firms like NEC, Toshiba, Fujitsu, Nissan, Yokogawa Electric, etc., decided to stay back in the region or after retirement, they, along with employees from the R&D divisions of these firms, started establishing their own small and mid-size processing firms in TAMA. Universities in the region also started to participate in the growth of the region — trained graduates found employment in these firms and professors were engaged in applied research supported by these firms. The focus, at this time, was to increase efficiency by agglomeration and formation of a network of producers. By mid-1990s, TAMA had emerged as the largest and strongest agglomeration of innovation-driven manufacturing firms in Japan (and perhaps in the world).

Kodama[5] and the TAMA Document[6] trace the findings of surveys done by the Kanto Region Bureau of The Ministry of International Trade and Industry (MITI) in 1996–1997, and the Research Institute of Economy, Trade and Industry (RIETI), Japan in 2003. As an outcome of the first survey of these firms, the formalization of the TAMA cluster was undertaken by the Japanese Ministry of Economy, Trade and Industry (METI). It was felt that the traditional system of mass production of large volume and standard products, in which the SMEs assembled products for large producers, had to be replaced by a more competitive model of linked SMEs that help produce a large variety of innovative products in small volumes. Different products, or variants and volumes, that a firm produced (i.e., flexible manufacturing), required linkages with different types of subcontractors/SMEs, with different kinds of competencies. This required flexible relationships based on needs and competencies, coordination between small companies, and above all, strong cooperation between university research and firms to rapidly manage advances in technology, and find solutions to customers' needs. This led to the formalization of cooperation in the region, through formation of TAMA Industrial Vitalization Association or popularly known as the TAMA Association,[7] in 1997. The purpose was to move the industrial agglomeration from mere production towards creation of new technologies — i.e., new products, processes and practices — and enterprises based on them, and use the network to produce these technologies competitively and locally. TAMA Association is the coordinating agency that brings together product-developing SMEs, product-processing SMEs,[8] large firms, local universities, local chambers of commerce, local municipalities, and the METI to revitalize Japanese manufacturing.

The DNA of TAMA comprises 'creation of new technologies, new products and new businesses by combination of different technologies and knowledge,[9] that is achieved through linkages between firms. The TAMA Network consists of R&D units of large enterprises like NEC, Hitachi, Yokogawa, Fuji Electric, etc. (about fifteen of them), 34 universities and colleges that have departments of sciences and faculties of engineering, about 300 product-developing SMEs, and over 16000 product processing SMEs. All are involved in R&D activities, although some may be in the product-development domain, while the others may be in the process-innovation domain. Different entities comprising this network play different roles. Large firms enhance the value of the cluster by providing access to new technologies, provide orders to other firms, and seek out cluster firms to meet various requirements. The product-developing SMEs serve the market needs through their own products; they focus on planning, design, final-stage assembly, testing, delivery and after-sales service to customers. They have a strong R&D orientation as reflected by the large number of patents filed, significant R&D expenditure, new product introductions, etc. (more on this later). They have a strong linkage with university research and researchers, to develop new technologies or solve specific engineering problems. In addition to being sources of new technology, the universities provide the valuable science and technology talent for firms in the region. The product-processing SMEs form the key manufacturing base for the product-developing SMEs. These processing firms produce parts (i.e., components/intermediates) according to designs provided to them by the product-developing SMEs. The precision level in their operations is very high. They excel in improving and developing new process technologies and they compete with each other for orders. The product-developing SMEs, on the other hand, rarely compete with one another as they are in different product segments. Instead, there is a great deal of collaboration on new designs and technology between them. This architecture gives the network enormous flexibility to produce a variety of products with original technologies, and helps in strengthening a wide base of processing competencies, thus enabling the network to become technologically very versatile. Last, but not the least, the TAMA Association forms the glue that keeps the network coordinated and facilitates continuous flow of innovation between firms, and from the firms to the market. Figure 6.1 shows the various elements of the TAMA network.

Figure 6.1 Various Elements of the TAMA Network

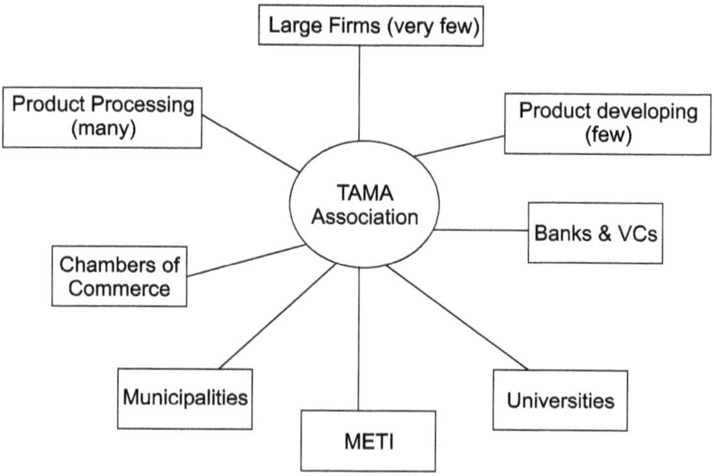

TAMA Association is the key lynchpin that helps reduce various market and capability failures for small firms in this region. In fact, it helps the network perform like a giant multi-product, multi-technology firm. Interestingly, the association cuts across multiple states, various regional cities, their municipalities (who support firms in their region through TAMA), firms producing variety of products, and institutions providing diverse research support. It has emerged as an organization that coordinates industry–university joint research, promotes inter-firm collaboration in research as well as orders and links SMEs with superior processing technologies to product-developing SMEs. TAMA secures active involvement of local governments, municipalities and chambers of commerce; manages assistance from METI to the cluster; and solves the diverse problems (technical, managerial or political) faced by it members. In addition, it provides a variety of services to its members:

1. *Provides a platform for financial support*: The Association works closely with local financial institutions in arranging loans for start-ups. Financial institutions seek help from TAMA Association for technical and market-based evaluation of loan requests by its firms. Experts from TAMA network help perform the evaluation using a business evaluation methodology developed by the Association. It also provides direct funding through two venture funds that were set up for start-ups (one is of 500 million Yen[10] while the other is of 1 billion Yen).
2. *Prepares local talent for joining the industrial workforce*: TAMA organizes job fairs for local students involving member firms, conducts a variety of skill-enhancing programmes for existing and fresh technicians, runs an internship service, etc.
3. *Provides new business and incubation support*: It runs a three-stage business plan competition with awards of USD 5000 to 10,000, leading to a matching session between Venture Capitalists (VCs) and the entrepreneurs. The Association has also developed a matching scheme — with the help of placement companies that are also members of TAMA Association — which helps staff from major companies to find positions in SMEs, thereby, often, facilitating a smooth transition for both large firms and employees facing closure or staff reduction. For example, in 2002, 25 firms requested the help of the Association in finding jobs for their employees. Firms recommended 65 employees. TAMA helped 12 of them find jobs, in ten firms, in functions like technology development, design (maximum number), sales, maintenance, R&D, Numerical Control (NC) programming

and accounting[11]). It also organizes an annual technology exchange for facilitating business linkage between local firms and large firms. It provides training to SMEs on presentation skills, and helps them with their presentations for this event. In 2003, local firms received 100 million Yen worth of orders during this event.

4. *Provides information network services*: TAMA, through its website, provides access to databases on firms and their capabilities, database on university facilities, faculty research and resources, helps companies set up their Information Technology (IT) based promotion environment as well as transaction systems like Electronic Data Interchange (EDI), disseminates information of interest to industry, and broadcasts events.
5. *Arranges technology exhibitions, disseminates research findings of universities, and provides overseas market support to its members:* TAMA has developed special exchange programmes with industry associations and firms in other countries. One such programme is with Italian SMEs enabling interaction between Italian and Japanese SMEs, especially, the design and technology firms and the universities.
6. *Arranges interaction between the technology and entrepreneur communities through 'Company Visits' and 'Mini-TAMA Meetings'*: It organizes visit of members to other firms in the cluster, in order to enhance learning and appreciation of difficulties. It also holds a monthly meeting (or Mini TAMA) to promote exchange of ideas and business between members.
7. Amongst the six hundred members of TAMA are 300 firms, 34 universities, individual professionals and professors, 78 banks and chambers of commerce, city councils (about 20), etc. Most employees of TAMA Association (five out of nine) have been loaned by various organizations, and they run the association on a full-time basis.

The success of the TAMA network can be gauged by processes that it is able to unleash, to integrate the capabilities of SMEs in the local economy and to reduce the market failures faced by small firms. In more quantitative terms, TAMA firms have exhibited excellent performance over other SMEs. For example, the number of new products introduced by TAMA firms in the market in the last three years was 22 as opposed to four by non-member SMEs; the percentage of firms collaborating with research universities was 65 which was an increase of 55 percent over the last five years; 69.2 percent of TAMA firms held patents as opposed to 29.6 percent of all SMEs in Japan; the R&D expenditure (per unit of sales) was five percent for product-developing SMEs and two percent for product-processing SMEs; and the profit to sales ratio was two percent for product-developing SMEs as compared to a median value of one percent for all Japan Machinery and Metal Manufacturing SMEs.[12] This data is from a survey done in the year 2001. Interestingly, according to the same survey, while the number of customers of product-developing SMEs who are members of TAMA was about the same as those who are not members of TAMA, TAMA product-processing SMEs had five times more customers than non-member product processing SMEs. All of the above point towards enhanced capabilities, and consequently, productivity of small firms when they become part of a collaborating network. The TAMA Network is an example of leadership role played by a coordinating agency, in bringing horizontal linkages between firms, with active support from local and federal government by building capability-enhancing infrastructure.

The Wenzhou Industrial Development Model in China

Wenzhou, a municipality located in the south-eastern coast of Zhejiang province of China, has become a model of transformation of a weak rural economy into an explosive global manufacturing centre, dominated by networks of tiny and small producers. In 2002, manufacturing accounted for 49.8 percent of Wenzhou's Gross Domestic Product (GDP). Its main products and market shares are given in the table below:

Table 6.2 Wenzhou's Market Share in China of Various Products

Product	*Market Share in China (percent)*
Metal Cigarette Lighter	90 (and 70 percent of world' reusable lighters)
Spectacle Frame	80
Locks	65
Razors	60
Plastic Products	56
Mechanical Pencils & Ball Pens	30
Footwear	20
Clothing	10
Low-Voltage Electric Appliances	70

Source: TDCTrade.Com, Barber, 2004

In 2002, the number of individually owned (or household) units and private firms stood at 202,458 and 28,430 respectively with a GDP of 106.1 billion Renminbi Yuan (RMB).[13] Of these, only 3700 firms had annual sales of five million RMB or more. The average number of persons working in the individually owned companies was 2.1 while that for private firms was 13 in 2002. Exports from Wenzhou were of the tune of USD 2.65 billion and imports were valued at USD 804 million.[14] This is impressive when compared with the fact that in 1970, the poverty level in Wenzhou was above China's average,[15] and its revenue from industrial and agricultural output was only 191 million Yuan.[16] The urban area of Wenzhou, about 488 square km., is barely four percent of the municipality's area. Rural incomes in Wenzhou have grown from 165 Yuan per person per year in 1978 (one of the lowest in the nation) to 924 Yuan per person per year in 1989 as compared to the national average of 601 Yuans.[17] This figure grew to 4683 Yuans for rural residents and 13,200 Yuans for urban residents in 2001; the average annual growth rate of GDP during 1978–2001 in Wenzhou has been 20.3 percent.[18]

Wenzhou is separated from other parts of China by a rugged mountainous territory on three sides and sea on the fourth, which had kept it in a bit of isolation from the rest of China. Absence of arable land led to the development of rural household industries, and consequently, local markets. Over ages, people from other parts of China have been using the isolated geography of Wenzhou to escape the civil wars of China as well as the 'move to farms' movement.[19] This increase in population put pressure on local resources and led to a heavy 'out-migration' from Wenzhou — many of these people carried products made in Wenzhou to markets outside thereby setting up a trading link, initially within China, and subsequently outside. This was the start of the formation of a supply chain. It brought goods from Wenzhou to the world and the feedback from customers on design and quality back to small producers in Wenzhou. It is interesting that until 1990, there was neither a highway nor railway leading to Wenzhou, and the city struggled to fight poverty after the revolution of 1949.[20] Today, the city is modern and is even privately funding the re-building of its international airport. It is useful to look at features of this transformation process that helped connect a large number of small producers to the market, and the process used to convert this weak sub-economy into an important industrial environment in China.

Characteristics of the Wenzhou Model of Industrial Development

a. Large number of small producers

Most producers in Wenzhou are household units producing a small range of products or parts therein. The large number of such producers gives the cluster a network economy of scale in the output. For instance, in 2001, there were 3000 individual and private enterprises in one of the clusters (in Liushi town) of Wenzhou employing 10,000 people (including the promoters), i.e., 3.3 employees per enterprise. Of these firms, 220 were International Organisation of Standardization (ISO) certified and 200 had UC&CE certifications. The annual trade volume was 10 billion Yuan, and it earned USD 0.3 billion in foreign exchange.

The cluster produced Industrial Electrical Appliances and was ranked first in China. Similarly, Tiger brand lighters (that has the highest market share in the world) are produced by over 300 manufacturers — producing over 0.5 billion lighters and a sales value of two billion Yuan (80 percent of production is exported). The plastic-weaving cluster in Wenzhou comprised 1600 enterprises in 2001, employed 42,000 people with an annual output value of 20 billion Yuan accounting for two-thirds of sales in the Chinese market. The lock cluster comprised 400 manufacturers in 2001 with a total output of five billion Yuan, representing 65 percent of the domestic market share. This cluster is the first in China in terms of market share and sold its products to 60 countries.[21]

Firms in Wenzhou have been relying primarily on family labor and most often, the house is the site of production. Most producers produce small parts (or 'small commodities') which are then assembled by other firms. The production structure is given in Figure 6.2.

Figure 6.2 Production Structure in Wenzhou

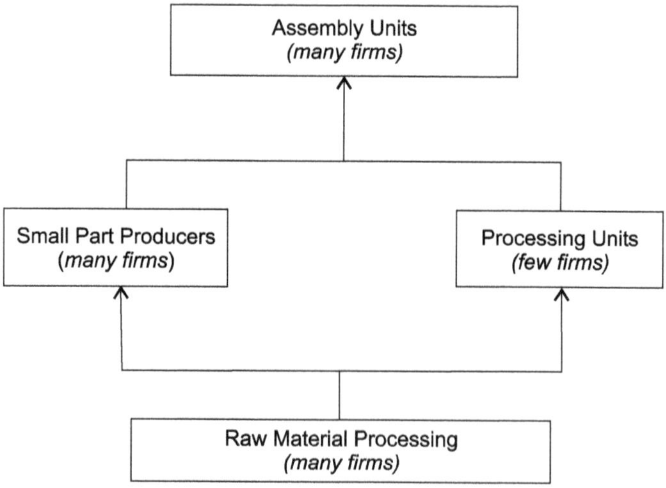

Production is mostly subcontracted by assembly or exporting firms who place orders, and often provide material either themselves or through specialized raw material procuring firms. Processing firms have special equipments for performing specialized processes on raw material or parts, e.g. electroplating. Sometimes they produce machinery locally. The key concept in this development has been 'one village one product' and 'one township one trade', thereby, getting the advantage of agglomeration economy while providing diversity in the product market. This has also helped reduce unnecessary competition between villages, thereby, conserving scarce resources. Often, parts are produced in villages while the assembly is done in larger market towns.

The structure of the industry has evolved in an interesting manner. Some household enterprises or individual private enterprises registered their business as a collective enterprise, when it really was privately managed and owned (especially in the 1970s and 1980s when private enterprise was forbidden in China. This phenomenon was termed as 'wearing a red hat'.[22] The collective enterprise received government support and escaped social and political ostracization for making profits, paid a fee to local party cadres for registration (often these firms were operated by cadres and their families), received bank loans and tax breaks, could lease land to others, etc. Similarly, there were 'hanger-on' firms — small private firms that provided subcontracting services to publicly owned firms. This allowed them to receive assistance such as raw material, use of publicly owned firm's resources, stationery, etc. They could also borrow official identification of the public firm, get legal status for transactions, use public firm's bank account to park money etc., all for a fee.[23] Such mechanisms also allowed 'interlocking control' by the collective in individual firms or by large public sector firms in small 'hanger-on' firms. Of late, many of the collective firms have turned into 'joint stock firms' with no government involvement but jointly owned by many. This helped firms to gradually accumulate capital and knowledge of production, and subsequently, expand their scale of operation.[24]

Subcontracting and the attendant division of labour has helped considerably in reducing the cost of production in Wenzhou. Clustering of firms in high concentration led to agglomeration economies in terms of scale of output from a single location, information on production technology and practices, subcontracting between small and specialized enterprises, and development of skilled labour market.[25] This has improved productivity, quality, and efficiency of operations and consequently the competitiveness of these small producers.

b. Specialized markets for products

Another unique aspect of the Wenzhou model is the establishment of, and access to local and specialized markets for parts and products. The common refrain of Wenzhou has been 'Small Commodities and Big Market'. Wenzhou municipality has set up 514 specialized markets whose business volume exceeds 47.5 billion Yuan — eight of these had turnover of 1 billion Yuan while more than 68 have business volumes greater than 100 Yuan.[26]

These markets ranged from village-and town-level fairs to permanent local wholesale markets, focusing on single or several related commodities. The market brings local parts-producers, assemblers, local selling agents, national buyers (and now international buyers) together. They set prices for the products, negotiate orders and lead times and book capacity, and provide valuable feedback on the quality and design of products as well as the delivery process. Another unusual aspect of these markets has been that they have created a market for parts produced by small household producers. Now, part-producers have direct access to buyers as well. Interestingly, small household producers have used 'mail orders' effectively to reach distant markets. These producers send catalogues and order forms to state and collective enterprises, department stores, and supply/marketing firms with information on their products and parts.[27]

c. Credit

In the early 1960s and 1970s, when people of Wenzhou started to move away from agriculture due to low productivity and shrinking incomes, they did not find the large industry ready to hire them. Most set up a household or a home firm to produce small goods of daily use that could be sold in the local market, and which was not produced by large firms. Initial capital was required for raw material as labour was coming from members of the family. This funding came largely from the family. Over the years, these entrepreneurs pooled their resources called the 'folk capital', to run their collective enterprise. Since 1949, Wenzhou has seen a growth of private 'money houses' or individual lenders (or 'specialized financial households') that has provided credit to small producers. They also set their interest rates—which was higher than state banks but most private producers were ineligible for loans from them in any case—with the knowledge of the local government. Collective firms and 'hanger-on' firms could borrow from the state banks. Even till 2001, less than one percent of loans from state banks went to private enterprises.[28] However, as business grew in Wenzhou, credit from non-government sources became plentiful and several new sources also emerged. For example, social organizations like Rural Cooperation Savings Foundation, Helping the Poor Foundation, etc., could collect funds as they paid higher interest rates to their subscribers than state banks. They, in turn, lent money to local enterprises of Wenzhou. After 1983, many farmers leased (or transferred) their agriculture land to others for a payment which provided the capital needed to invest in small manufacturing enterprises. Growth in manufacturing led to gradual accumulation of capital. Only recently has Wenzhou seen foreign investment in its enterprises. For example, the year 2002 saw implementation of about 46 FDI projects worth about USD 80 million. By 2003, observing the success of Wenzhou, many banks made it a priority to lend to the small producers of Wenzhou.

d. Sales network

In the early days of the cluster, most individual household firms relied on peasant traders to carry their goods from one village to another or to another part of China. Since Wenzhou did not have strong agriculture and industry, many people migrated to other parts of China, and subsequently, to Europe and the United

States. These people carried goods from Wenzhou that they sold elsewhere to survive since travel to another city was often not authorized by a 'permit' or by a work unit. This was the start of the formation of a sales network for enterprises of Wenzhou.

Since many of the products were small in size, low tech, and for daily use, producers found a great domestic demand for them in the 1970s and 1980s, after years of shortage in the command and control economy of China.[29] Several local sales agents emerged. They helped organize roadside markets (in the initial years of liberalizations), took local consumer and industrial products to stores in other parts of China, and subsequently, helped bring customers to the village trade fairs. These trade fairs also gave opportunity to outside traders and merchants to source goods for their international clients from Wenzhou. This helped coordinate the linkage between the final buyers and the producers. Many of these were former citizens of Wenzhou who had migrated to other countries. In recent times, 83 percent of immigrants from Zhejiang province to Europe have been from Wenzhou, while the annual number of people migrating to other parts of China from Wenzhou is estimated at 270,000.[30] These immigrants helped bigger entrepreneurs to register companies outside China, especially in Bermuda, and raise equity on international stock exchanges.

In early 2000, it was estimated that there were over 10,000 chain stores all over China that were selling only Wenzhou garments.[31] Migrant sales agents from Wenzhou not only provided valuable information on outside markets but also performed the contracting function for these products. Often they provided the capital needed to produce these products. Many of them brought back information on production technology that was used elsewhere. The agents consolidated the important logistics function for the small producers.

e. Government as a facilitator and coordinator

Since 1949, the policy of 'no undue interference from outside' by local government officials has helped facilitate the spread of small businesses in Wenzhou. Household enterprises could be kept operating quietly, and could be controlled easily at a time when private enterprise was not to be allowed. In that, the interest of the peasants coincided with those of the cadre. Often private business contributed towards public infrastructure like providing offices for party cadres. Several members of the party or their spouses would own household enterprises. Many joined these enterprises after leaving government jobs. State enterprises also subcontracted jobs to smaller firms.

The local government was proactive in enhancing the status of Wenzhou economy and the enterprises. Many of the fairs/markets were organized and supported by them. It periodically cracks down on counterfeits or products of low quality that are found in Wenzhou markets. In one instance, it set up centres on highways and ports, to prevent such products from leaving Wenzhou.

In recent times, the local government has invested in quality-testing centres, training programmes, improvements in housing standards, facilitated resident registration procedures, and has argued for higher salaries to attract talent to Wenzhou.[32] The local government of Liushi city in Wenzhou, for example, even passed laws to ban certain kinds of technology in the low voltage equipment cluster, in order to force investment in newer process technologies and thereby improve efficiency of operations.[33] Local government has taken a keen interest in the coordination of activities ranging from sheltering small enterprises, organizing local markets, facilitating credit, and in more recent times, bringing investments in infrastructure like highways, railways, airports, and ports.

Stages of Growth in Wenzhou

The trajectory of development of enterprise in Wenzhou can be categorized in the following three stages:

1. **Stage I (1970s and early 1980s):** This phase marks the development of local markets for simple imitative products for daily need. These were sold mostly within Wenzhou though some merchants carried these products from roadside markets to other places. These were characterized as low-cost and low-quality products.

2. **Stage II (1980s and early 1990s):** The number of households producing simple commodities increased in this period. Specialized markets emerge in Wenzhou. The volume of output from the network of producers grew considerably. Wenzhou goods were now being increasingly sold in national markets. Products were still low cost and of low quality. Sale of products took place mostly through Wenzhou traders (26.5 percent of total sales), local wholesale markets (23.5 percent), sales agencies (22 percent) and own retail outlets (9.5 percent).[34] Interestingly, most of the part producers had an arm's length relationship with assemblers. This stage was also characterized by early improvements in quality of products through introduction of inspection — though processes remained, by and large, old — by a few industry leaders.
3. **Stage III (1990 onwards):** The last decade has seen a variety of changes in the industry both in terms of its structure as well as value added. Sonobe et al.[35] surveyed firms in Wenzhou and found an increase in number of enterprises — though the number of independent enterprises initially increased and then decreased — value added, sales revenue, number of employees per firm, and capital stock of firms between 1999 and 2000. As incomes in China grew, demand for quality products also increased. Larger firms installed quality systems, predominantly, inspection-based, though there was no substantial change in production technology. However, a few independent firms broke ranks on technology, and are emerging as world leaders in their specific product domain. For example, one such group called 'Chint Group' which manufactures low and medium-voltage electrical equipment was started by setting up a household firm twenty years ago, and now employs 13,000 workers in more than 50 factories. It has sales of USD 2 billion and one of its factories employs Japanese robots for pouring molten plastic around the clock — the factory produces electrical switches on several manual assembly lines employing workers from all over China at an average monthly wage of USD 240.[36]

The tools of production have started to change in the last decade even for other smaller producers. Products still remain low cost but of improved quality, and are produced in large volumes from the cluster. There is evidence of merger of smaller firms into large groups. In the firms sampled by Sonobe et al.,[37] the largest group had 70 subsidiaries in 2000. They also found that several small firms exited manufacturing, as their operations were inefficient compared to their competitors in the village/township cluster.

Incidentally, this period also saw an increase in export-led growth through Wenzhou merchants overseas, and recently through investment by some foreign firms. Consistent with the evolving strategy to increase value added in the cluster, lead firms in Wenzhou developed their own brands to distinguish their quality — e.g., Chint and Delixi are now China's largest producers, and most well-known brands in low and medium-voltage appliances; Tiger brand of lighters is known worldwide; Kanghai, Jierda, Aokang, and Dongyi brands have made Wenzhou the shoe capital of China; Baoxiniao and Zhuangji are known garment brands in China, etc. The cluster has also seen changes in the downstream part of its supply chain. It distributes more than 50 percent of its goods through sales agencies as opposed to 22 percent ten years ago, and more than 27 percent through its own retail outlets.[38] This has been a major shift in the 2000s — moving away from local wholesale markets and Wenzhou traders. Large firms are setting up their own stores all over China. There is also evidence of substantial effort to seek national and international certification of the production facilities and products,[39] which reflects an increasing level of awareness of quality and its value.

In summary, the Wenzhou model of development has moved from being a 'merchant-led' industrial development model (as termed by Sonobe et al.[40]) where low quality, low cost products were produced by household enterprises and sold in specialized local markets by producers and elsewhere by Wenzhou traders, to becoming a 'manufacturing-driven' industrial cluster with agglomeration economies, emergence of larger firms through merger of small producers, and capacity expansion and distribution through sales agents and own retail networks. It has transformed a poor and rural municipality of Wenzhou into a vibrant and more affluent industrial economy whose spirit can be captured by the terms 'for a penny' and 'to earn a little money',[41] implying work for small margins, to earn whatever one can, and produce large volumes by getting specialized.

The Engineering Network of Rajkot in India[42]

The Rajkot Engineering Cluster is an interesting comparison to the organizational forms seen in TAMA and Wenzhou. Rajkot, an industrial town, is located in Gujarat in Western India, about 250 km south-west of Ahmedabad and about 600 km north of Mumbai. Rajkot lies in the Saurashtra region of Gujarat which is known for its engineering craftsmanship. The engineering industry of Saurashtra (which has industrial cities like Rajkot, Jamnagar, Surendranagar, etc.) grew with the arrival of craftsmen from Pakistan after the Partition of India in 1947. Since then, the industry has grown manifold and developed strengths, to integrate small processors in an industrial model of assembly by larger firms, and production of parts (including specialized processing) by small firms.

The Rajkot cluster started by producing small items in brass tumblers, cups, bells and other similar items for home use, and small industrial parts as substitutes for imported items. This required casting and machining of small brass parts. The industry grew by leaps and bounds with the Government of India announcing subsidies in mid-1960s for low speed, low horsepower diesel engines for implementing its green revolution in agriculture, especially in rural India. This sector was reserved for small scale industry (SSI) by the government, which was keen on indigenizing a largely imported product. Small firms in Rajkot became the hub of low-speed diesel engine manufacturing in India. The customers were chiefly small and mid-size farmers in rural India who required a low-cost option for driving irrigation.

At its peak in the mid-1990s, there were about 8000 factories employing over 500,000 people that produced about 650,000 diesel engines of 10 horse power (hp). Annual sales turnover of the industry was about ₹ 250 crores.[43] Rajkot accounted for about 60 percent of the country's total production and about half of India's export of these types of diesel engines. This production was supported by about 500 foundry units. Most of the diesel engine and foundry firms were in the SME sector. However, a few large manufacturing and assembly firms like Field Marshall, K. Rasiklal, Ashok, etc., had close to fifty percent share of the Rajkot market and would set market prices. Since a diesel engine comprises about 350 parts, a large majority of SMEs either produced parts or performed specialized processing job-work like finishing, machining, drilling, etc., for assemblers and part producers. The rise of the engineering cluster led to a growth of a machine tool industry in Rajkot which catered to local needs. Most firms manufactured conventional machine tools like, lathes, drilling machines, cutting machines, etc. while a few firms produced computer numerical control (CNC) machines for local use.

However, by 2001, the number of diesel engine sets sold fell to 350,000, and by 2005, this number further reduced to about 100,000. Rajkot's share in export fell to 10 percent in 2005. In 2005, only 100 diesel engine manufacturers and about 4000 auxiliary units remained open of which more than half were on the verge of closure.[44] What led to a virtual collapse of a vibrant industrial cluster is an interesting case study.

Some Salient Characteristics of the Rajkot Cluster

1. **Policy-induced cluster:** The growth and subsequent decline of the Rajkot Engineering Cluster appears to be a response to government's policy regarding diesel engines. The government decided to promote the use of low-speed, low-horsepower diesel engine for introducing automation in irrigation of fields during the agriculture revolution of the 1960s. The demand for low-speed diesel engines grew tremendously as the government provided subsidy to farmers for its purchase.

This growth attracted new entrepreneurs to establish manufacturing business. Since the production of diesel engines and its parts was reserved for SSIs, firms remained small. Whenever the business of a producer grew, a new firm was established at the same location or elsewhere in Rajkot, as capacity expansion would send their investments outside the prescribed range, and make it ineligible for being an SSI and consequently, the accompanying benefits. It can be said that Rajkot producers locked themselves, technologically and product mix-wise, to less than 10 hp diesel engine production. In the early 1990s, the government decided to remove the subsidy on purchase of low-speed diesel engines by farmers. It also imposed excise duty on diesel engines. By this time, the power situation in the country had also changed considerably, and farmers were starting to purchase better-quality technology rotary engines. The demand for low-speed diesel engines came down dramatically in the decade following these developments. By early 2000, the industry structure had changed in Rajkot; a number of small firms had closed down while others started to diversify into new manufacturing opportunities.

2. **Intense outsourcing** or Linkages between firms in Rajkot: Two features of the industry have contributed significantly to the formation of strong and active linkages between small firms in Rajkot: (a) diesel engine manufacturing requires fabricating 350–400 components and then assembling these parts, and (b) reservation of diesel engine production for small scale industries. While the former meant deriving economical relationship between assemblers and producers of parts, as the product was amenable to component production in batches and assembly in small to mid volumes. The latter meant that firms had to remain small even if there was growth in demand for the product, in order to be eligible to produce the product as well as to get various tax-related benefits. Another related reason was that small firms did not have the resources needed to integrate vertically. Hence, they would depend on other firms to supply them with components

and services to complete an order. This led to the growth of 'processing firms', firms that would do rough casting and finish, machining, drawing, etc., firms performing individual operations for other firms similar to the 'product-processing firms' of TAMA. Basant's[45] survey of firms in 1997 revealed that about 77 percent of sample firms outsourced jobs to other firms in Rajkot. Amongst others, the benefits cited were ability to meet orders from premises of limited size, ability to reduce costs; however, it led to intense price-based competition between assemblers and subcontractors alike etc. Often the outsourced firms were owned by family, friends or former employees. This was both good and bad — while negotiations on price, or delivery dates, or expediting an order were facilitated, sometimes this close relationship came in the way of business decisions especially those relating to quality or meeting delivery times.

Along with outsourcing of orders, one finds outsourcing of labour to be quite prevalent in Rajkot—contract workers who may perform tasks like painting, packing, casting and even assembly. Often these contract workers take complete responsibility of an operation or a section, get incentives for producing more, shift from working on one operation to another based on the requirement (i.e., provide flexibility) and work in the same firm for several years.

In Rajkot, it has also not been uncommon for employees to set up their own small enterprise and become vendors to their former employers. This has helped in minimizing the loss of skills that accompanies the turnover of employees. At the same time, it has assisted in the listing of suppliers' and strengthening of linkages — a 'deepening' of the Rajkot network. While availability of labour is not a problem, absenteeism is. Workers tend to take leave during the sowing and harvesting seasons, as they return to their villages during these periods to work on their farms.

There is intense price competition amongst producers or processors. As a result, intense cooperation has been missing amongst them. Trust is missing, and firms would not share orders for fear of losing a customer to a competitor. The Rajkot Engineering Association (REA) has been trying to collectively enhance the capabilities of the firms.

3. **Technology and Productivity:** The 'Rajkot engines' — 'Lister' and 'Petter' types — are based on technology that is over a hundred years old. They are low (Lister type) to medium speed (Petter type), low horsepower (less than ten hp) and have high weight to power ratios (hence, fuel inefficient). They also have limited application, restricted to agriculture. Product technology, consequently, was outdated with little scope for innovation in its design. Product design and the application domain also imposed constraints on process technology. As an aside, free power and water to farmers (as part of strategy by political parties in India to gain popular support and votes during elections) also ensured that more efficient technology was not imperative for rural farmers. The drivers for product innovation were weak and over the years, the product did not change.

Since the product mix was low-tech and did not change over the years, capabilities of firms also did not change significantly. With government subsidy, firms could sell and survive. Some firms, needless to say, broke ranks and entered new areas or developed newer technologies and competencies. As Basant[46] has observed, most entrepreneurs did not have formal engineering education but had picked up skills by working and running their plants. Such a strategy worked well in the initial years; supported by government subsidies for diesel engines, sales were high and technology development or productivity improvements were non-existent. Basant's survey reveals high defect rates amongst sample firms, insignificant design modification, low quality of casting, usage of low-quality processing equipment etc. Capabilities that were considered adequate for one product-market segment were inadequate for another. When one of the biggest firms, Field Marshall, tried to produce better performing, higher quality HATZ diesel engines for a German firm, they failed to meet the quality requirements.[47] The average capabilities of vendors in Rajkot cluster were not sufficient to meet more stringent requirements of casting, finishing etc. Moreover, with the small volumes of HATZ engine, Field Marshall was not successful in getting the vendors to invest in newer technology and consequently, improve their capabilities. More than 50 diesel engine units shut down in the decade of 1990s. Since the Rajkot cluster comprises predominantly of processing firms (firms that perform a specialized processes like casting, rough finishing, wire drawing, metal removal, lathe operations, etc.) for a manufacturer of diesel engines, it was essential for new technology or good manufacturing practices to spread throughout the cluster for the quality of the product to improve. This did not happen in Rajkot. The capabilities of a majority of firms did not change substantially. By 2005, barely 20 percent of the 400 foundries were doing well while the rest were struggling to survive.

On the behavioural side, the cluster suffered due to poor cooperation. During our interviews, several entrepreneurs mentioned 'lack of trust' as a key reason behind low cooperation. It appears that over the years, competition between an increasing number of small producers had become very intense. Since the market for diesel engines was not expanding, firms became very secretive, did not invest much in training from a fear of losing trained workers to a competitor, and became technologically and socially very conservative. There was hardly any exchange of ideas between firms, and with the decline in the slow speed diesel engine industry, a number of small processors started closing down their plants. The only time they came together, at the initiative of the Rajkot Engineering Association, was to jointly purchase coal and pig iron for their foundries.

While the cluster was declining, a small set of producers (about 20 percent of the 1,100 members of the Rajkot Engineering Association) led by firms like Patel Brass Works (PBW) were slowly diversifying into auto-components, motors, machine tools, bearings, etc., that used their existing engineering capabilities. They saw diversification as the only survival strategy. These included some foundries and processing units. The product-producing firms also sought original equipment manufacturer (OEM) customers and started branding for the replacement market. They invested in newer technologies, entered newer product domains with old skills, and newer strategies for both domestic and global markets. Rajkot is undergoing a major restructuring not only in terms of the industrial structure but also in products and technologies. For example, firms have understood the need to upgrade the tools of productions in order to achieve higher quality. One example is that almost all these small firms have introduced CNC machines on their shop floor; this has also opened a new market for CNC-based machine tools for the machine tool producers of Rajkot. Similarly, auto-forging plants have been set up by some of these SMEs thereby changing the capability environment of the cluster. Larger SMEs like PBW made investment in their exclusive vendors for upgradation of technology.[48]

In this context, the trajectory of growth of two SMEs of Rajkot, PBW and MacPower CNC, is instructive. PBW started by producing brass tumblers and bearings for diesel engines. It understood the market dynamics of Rajkot and diversified into bearings for critical applications in turbines, compressors, etc. It has patented and improved upon the proprietary technology of centrifugal casting, modernized its shop floor, developed a catalogue of over 2000 compressor parts that it can produce and deliver within 48 hours anywhere in the world, developed capabilities to produce over 4000 bearings for various industrial applications (including bi-metal and tri-metal bearings), has developed a sophisticated testing facility with state-of-art equipments, and has a work force of over 300 people. PBW, now, has a turnover of more than ₹ 30 crore with customers all over the world including General Motors, GE in The United States and in Japan.

Similar to PBW is the story of MacPower CNC. It started designing traditional lathes for producers of Rajkot; many of these firms operated with workers that had no formal education. Today, it is one of the most prominent CNC machine-tool producers of Rajkot with a turnover reaching ₹ 100 crore. Along with a couple of other similar firms, it has helped Rajkot become the largest producer of conventional machine tools in India, and is the third largest cluster of CNC machine tools in India, and is poised to soon overtake the second largest cluster of Pune. MacPower competes with products from abroad that are imported in the Indian market. It designs, fine finishes and assembles its machines in-house while sub-contracting other operations to some of the processing SMEs of Rajkot.

The metal processing industry was the worst hit with the decline of diesel-engine manufacturing in Rajkot. After struggling to survive for a decade, in 2000, they started looking for new markets for their skills in engineering. The auto industry in India started to move into growth phase with the ramp up of auto production (including two wheelers) and the entry of several global producers. The number of units producing auto-components rose from five in 2002 to 150 in 2005, and the industry has been growing at the rate of 25 percent per annum. Their turnover has grown from ₹ 100 crore in 2000 to ₹ 1000 crore in 2005.[49] The rate of export orders have also increased from 1.5 percent to 20 percent of the total value within a year. Amongst their customers are almost all the auto and tractor producers that are operating in India.

While the formation of the diesel-engine-oriented manufacturing was induced by policy, the entry of some of these firms into other sectors—due to the decline of the diesel engine industry—has been driven by the conditions of the market. Technological capabilities that were built over the earlier decades formed the basis for diversification to new industrial sectors. Rajkot is a good example of how SMEs who build capabilities, can deploy them flexibly into new application domains as the market environment changes. Excelling in these newer segments would require building on these skills and acquiring newer ones in the future.

The Three Networks: Approximations to a Generic Model

The distinctive strengths of the TAMA and the Wenzhou networks are quite apparent. The resilience of the Rajkot firms speaks highly of its accumulation of technological capabilities that has enabled firms to enter into newer sectors. Table 6.3 gives a comparative picture of the three networks, i.e., TAMA, Wenzhou and Rajkot, over some key dimensions. Each of the three clusters comprises a variety of different kinds of firms that were playing a distinctive role. TAMA's ability to draw processing firms to develop technological competencies for developing new products stands out. Wenzhou's ability to coordinate across a large number of small firms and to standardize processes in order to improve quality is noticeable. While both of the above had support of the local government, the role played by the government in Rajkot was quite distinctive, which was the cause of its growth as well as its decline. Nevertheless, they make a strong case for collaboration amongst SMEs to form a dynamic network. It must be mentioned that in all the three cases, there is less direct interaction with large firms, though their technology-driving and demand-generating roles cannot be ignored. While markets for the product-developing firms may have been away from the respective cluster, market for the processing firms as well as part producers was quite local. The TAMA and Wenzhou clusters produced a large variety of products at small volumes — a typical job-shop-like characteristic. Rajkot's decline in the diesel engine business also points towards lack of variety as a cause. Process innovation and the role of technology-driven intermediaries is also seen across the cases though their strategic roles may have varied. These intermediaries planned their role. TAMA Association is the most prominent and visible coordinator. Based on structures, processes, linkages, and technologies found in these clusters, we now present some generic observations towards building a network model of collaboration for small producers.

There appear to be six distinctive determinants of a collaborative model for engaging SMEs, in technological innovation, over a period of time. These are: Focus of the Firm, Interactive Producers, Processing and Product Manufacturing, Innovation Investment, Markets, Market Makers (and market making processes), and Regulatory Support.

Focus of the Firm: SMEs thrive best in an innovative environment as they can provide flexibility at least cost (more of job-shop-type producers). They are also able to switch over to new products with shifts in demand with much lower penalties as compared to large producers. This also means that SMEs need

Table 6.3 Comparison of the Three Models on Various Dimensions

	Nature of Firms	*Product Characteristics*	*Markets*	*Key Intermediaries*	*Supply Chain Linkages*	*Technology Linkages*	*Innovation Pattern*	*Strategic Focus*
TAMA	• Processing • Product Developing • Product firms differentiate by product	• Large Variety • Small to Mid Volume • Innovative Products	• National-Global • Market for End Products • Market for Processes	• TAMA Officials • Large Firms	• Large Producers • Other SMEs • Collaborative Growth	• Universities • Processing Firms • R&D Units of Large Firms	• Process Innovation • New Products	• Develop New Products through New technology
Wenzhou	• Product Developing • Processing • Part Manufacturing • Large number of firms producing the same end product	• Large Variety • Large volume for cluster (small for each firm) • Old Product-New Product market	• National-Global • Market for Parts • Market for End Products • Local Fairs & Selling Centres	• Local Agents • National and International Agents • Party Officials	• Alliance with Marketing Firms • Local and National Stores • Strong Collaboration between Producers	• Between SMEs	• Innovative Markets • Distributed Processing	• Develop Products for New Markets through distributed manufactur-ing — large volumes through large number of firms

(*Continued . . .*)

(Table 6.3 Continued . . .)

	Nature of Firms	*Product Characteristics*	*Markets*	*Key Intermediaries*	*Supply Chain Linkages*	*Technology Linkages*	*Innovation Pattern*	*Strategic Focus*
Rajkot	• Processing • Part Assembly • Many firms producing the same end product	• Standard Product • Mid Volume • Old (and Standard) Product	• National • Parts Subcon- tracted • Market for End Products	• Assemblers	• Assemblers • National Dealers • Poor Collabora- tion between Producers	• Between SMEs • Technical Centre	• Some Process Innovation – specific to the product	• Develop stan- dard product for local markets at lowest cost

to develop focused capabilities. This focus, in addition, forms a strong basis of collaboration between other firms in the cluster. It also allows the development of an eco-system with a breadth of capabilities that would be available in the network. This has a significant demand-enhancing effect — customers tend to prefer networks where they can source all their requirements, thereby reducing the need for information search, and consequently higher transactions cost. Firms get distinguished by distinctive capabilities that allow them to compete successfully in the market. Small producers are more vulnerable to price-based competition, hence flexibility allows them to survive better in markets where presence of 'similar' competitors helps a buyer or an agent, to negotiate the price downward.

SMEs are also more willing to reduce the price to win an order — only to find their inability to supply at the reduced price hence the temptation to reduce quality — rather than offer higher value product through flexibility, and quality of design and processes.

Interactive Producers: Small producers who either share orders (over and above their production capacities), or have an outsourcing relationship, or are part of a sales network (or umbrella branding), create strong synergies for sustaining growth. It ameliorates both, market failures as well as capabilities failure, in a cluster is small. While the former is recognized well in literature and practice, the latter has not been emphasized. It is the inability of small firms to constantly upgrade their technological and managerial capabilities — i.e., either due to isolation, or 'single-manager congestion' effect, or lack of opportunities, and high fixed costs for doing the same — that renders them less innovative and less effective, given their size. Interactive producers also broaden the market for skills and capabilities.

Processing and Product Manufacturing: Small firms need to distinguish between those that would develop an end product or even a component versus those that could excel in specialized processes like anodizing, welding, embroidery, precision drawing, precision milling, die making, etc. While the competitive requirements of a product-developing firm are well understood, those for the processing firms are not very clear in the minds of most producers. The need for capability enhancement and R&D in processing firms can yield significant returns; proprietary technology helps build natural barriers to competition as well as through intellectual property rights, and delivers a competitive edge that would win orders. Processing firms provide product manufacturers an opportunity to invest in product-enhancing technologies, rather than spreading their precious resource over processing technology. This has a capability-spreading effect and makes the cluster move towards higher capabilities and more profitable markets. The presence of a strong and a large number of processing firms in a cluster, also allows the product firms to be more flexible in changing their product mix as demand changes, due to access to diverse capabilities in developing and processing new products.

Innovation Investment: An ability to change the product mix requires investment in the development of distinctive capabilities and its continuous upgradation. While fixed costs deter producers from investing in capability-enhancing or quality-enhancing training as well as in equipment and R&D, common facilities on a pay and use basis, or service firms that help spread the overhead across many such firms, help improve capabilities considerably. For instance, a small producer may not be able to hire a Quality Consultant (who may need a minimum number of hours of consulting), but several producers could jointly do the same where each gets the services for a part of the contracted time). Linkages with universities are extremely valuable here. Such a shared investment in innovative processes also acts as an incentive to change more frequently and keep up with any changes in the market.

Markets: The key to successful integration of small producers in a growing network is to develop a 'market' for parts, components, and processing services. While this is related to subcontracting (with or without supply of blueprints and raw materials), it allows small firms to link up with other small intermediate producers, thereby widening (or broadening) the market. This diversification has two benefits: it creates opportunities for a range of producers, and it also allows tiny producers to enter the market domain by producing a small element of the end product. This effect is further intensified for processing firms. There are three things that this market needs to signal for it to become 'valid'. It must be of credible size for buyers to show interest (i.e., through a large number of producers either producing a wide variety or volumes that may help reduce the search/transaction cost of the buyer), reflect desirable quality and price (i.e., through effective operations and logistics management), and periodically bring newer upgrades on their parts or processes that add value to the buyer and the products that she produces (i.e., through innovation and flexibility of a job shop). Markets become the funnel for focusing buyer and seller's attention. A 'wide' market (one that reflects variety) also reduces the commitment that each seller has to make in producing goods upfront in an uncertain market. One probably might be able to argue that this widening could be uncertainty-reducing as well. Such markets favour the SME producer.

The question, of course, remains as to who will bring these part producers, or processing firms, together. While it is tempting to hypothesize that the market will bring them together, small firms face information asymmetry and hence may not be able to predict needs a priori, or may want to establish a unit once there is an evidence of demand. Hence, the need for a coordinating agency — individuals, government bridge-programmes, associations, or private firms that specialize in this particular activity.

Market Makers (and market-making processes): Market intermediaries perform many functions of which the following are extremely useful for SME development: making private information public, helping small producers develop market-driven capabilities, and provide a platform for SMEs to collaborate on common infrastructure. In the first role, market intermediaries like agents, wholesalers, retailers or even the government, bring producers' capabilities to buyers and buyer's requirements to producers. The State, especially in emerging markets provides the infrastructure — both physical and informational through the internet — for markets to be identified. Examples being Special Economic Zones (SEZs), regional produce markets, bazaars and *haats*, hawker centres, craft markets, etc. Recent experiments of

ITC eChaupal in India is a good example of private initiative in the same direction. Similarly, intermediaries like TAMA help SMEs develop specialized innovative capabilities, by linking small producers to R&D firms and University laboratories. TAMA meetings are mechanisms used to bring them together. Industry Associations could also perform the same task, of enhancing the capabilities of specific firms, by exposing them to new technologies, managerial practices, and specialized training. There are also examples of intermediaries existing as independent firms that provide services (e.g., logistics) whose absence may cause market failure for small producers. Credit Unions run by a group of producers or agencies for purchase of raw materials are quite common. Japanese automobile firms have often formed an association of their small suppliers (Supplier Association) to assist in their development. Sometimes, agents play the role of subcontractors that provide credit for production of goods by a network of producers. In summary, their role is one of keeping a set of diverse specialized firms integrated and independent.

Effective Market Coordinators—either individuals, institutions or firms—have often been the difference between successful clusters of small producers and a collection of struggling SMEs. They bring producers together in a network by leveraging the strengths of the group, and thereby helping the network become diverse and attractive to customers. Operationally, they assist in the pooling of diverse resources which allows small firms to manage uncertainty better. The motivation of the intermediaries, often, varies ranging from seeing economy grow to technology push to profits. Two interesting questions arise in this context: how to keep the market makers engaged with low, short term benefits, and distinguishing the role of the coordinator at the time of start-up of the network, and then through its growth phase (i.e., manage the transition from variety to volumes).

Regulatory Support: Focused support from the government helps reduce the market and capability failures would, in turn, deliver maximal benefits to small producers. These can be categorized as follows:

1. *Incorporation and Closure Support*: There is a need for processes that will facilitate setting up a small firm within a few days and at a low incorporation fee. Same for closure especially when the number of people working in the enterprise is low. This will allow risk-taking and encourage firms to invest in innovative activities. Self-reporting and high penalties for violation of norms, will address the monitoring issue without expending high overheads. Local industry associations and 'better business bureaus' are best suited for providing potential customers with an objective evaluation of the business practices of the small firm. This kind of intermediation requires professional management of these associations rather than government intervention.
2. *Access to Markets and Market Information*: Organized markets in the form of clusters, SEZs or local *haats* (weekly bazaars) etc., provide the necessary access to local markets. The only question that remains is, who sets them up and who manages them? Ideally a 'sellers association' would be most suited for running such an organization, and would require a private-enterprise approach to running the markets. Alternatively, an independent market-making firm might coordinate this activity between the buyers and the sellers. Buyers also find this mode of identification of markets easy. Liberal resources for participating in national and international fairs provide the necessary opportunities to identify external linkages. There appears to be a need for development of strong product-trading houses (some do exist in the garment sector in India, for instance). So, the regulatory role becomes one of assisting in the formation of the supply chain, and in ensuring that all elements of the chain are present to form an eco-system. Once it gets formed, local institutions take over to manage its development and growth.
3. *Market for Credit*: While informal mechanisms do develop for funding innovative enterprises, formal channels require arrangements that lead to lending for capital as well as working-capital expenses. Funding to support prototype

building activities help promote innovation amongst small producers. The market for this kind of funding is most underdeveloped; same is true for other 'epoch funding' requirements—funds needed at crucial junctures of the development of an SME. As a result, regulatory support that helps link the SME with credit guarantee trusts, venture capitalists, large producers, or banks supportive of SMEs becomes crucial.

4. *Capability-Building Support*: This remains the key differentiator between SME programmes that have succeeded and those that have not. Three elements of capability building requires attention:
   (i) Development of skills of both entrepreneurs (or potential entrepreneurs) as well as of employees of SMEs: this would include continuous training on new technology and processes, skills for managing a business including lean manufacturing practices, seeking funds and evaluating risks, processing market information and accessing new sources of business, innovation and new application domains, etc.
   (ii) Formation of linkages: it has been seen that firms that expend effort in forming linkages—or are part of a larger network—with customers, technology and material suppliers, research institutions, consultants, etc., develop strong capabilities in the product, process and practice domains.[50] Firms learn from these linkages and develop distinctive capabilities to compete in the market.
   (iii) *Continuous Upgradation of Tools of Production:* SMEs often suffer in terms of quality and productivity due to a prolonged use of inappropriate tools of production. For example, most small producers in the machine shops of Taiwan operate CNC machines, while those in India will use 'addas' or manual lathes that require enormous operator effort, have short life, and are not capable of producing precision products. This prevents the small producers from playing up their strengths of a low

volume, high variety, and value producer. Similarly, use of a pneumatic wrench controls the torque better and renders an application (e.g., wheel-fitting in an automotive vehicle) more safe and conforming to precise specifications. Use of manual wrench, in this case, could lead to inappropriate application of torque, hence may lead to unsafe usage of a machine, and decreases operator productivity. In the long run, this increases the cost of operations.

Policies that strengthen the above or provide incentives for the same, help small producers build distinctive capabilities and become innovative.

Needless to mention, all other infrastructure support (i.e., road, power, water, etc.) that are required by any other firm would be required by small producers as well. In essence, SME-support programmes should try to reduce the risk of operations for a small producer, thereby increasing the probability that a small producer will continuously invest in the enterprise.

Conclusion

Small producers have an inherent advantage in performing innovative activities within their organization. However, they also face coordination failures due to a variety of reasons. To overcome these disadvantages, firms need to design ways and means of cooperation, especially on those issues that cause market and capabilities failure.

In this paper, we have presented learning from 3 global networks that have adopted a variety of mechanisms of coordi-nation between small producers that has led to both capability enhancement and demand enhancement. We argue that capability enhancement effects play as significant a role as demand enhancement effects in the growth of small firms. Coordination that allows firms to improve their capabilities, enhances productivity as well as innovative capabilities to develop new products and processes. Here the focus of the firm, i.e., its key line of business, its recognition of its variety-producing abilities, understanding its role as a product-developing firm or a product-processing firm, etc., and its linkages with key suppliers (both of raw material and equipment) and customers, become extremely crucial. Firms learn from each other through consultants, suppliers, customers, employees who transit from one firm to another, etc., thereby influencing the productivity and innovation regimes within the firm. On the other hand, cooperation within a network enhances the demand faced by the entire network, as potential customers are assured of the availability of various price-technology-quality options, a large variety of product offerings at the same location, and a higher opportunity to experiment with new designs, new materials, etc. The key to success in this structuring is the extent of trust and reliability within the members of the network. Many firms have failed to cooperate because of their inability to trust others or forge trustworthy partnerships. Once again, individuals often emerge as role models in the local eco-system who play the coordinator's role and make the market for network operations. It should not be construed that this coordination is limiting in competition—anecdotal evidence from the field is just the opposite; customers are able to create intense competition between small producers when they are located in a cluster due to access to more number of firms. Another aspect relates to creation of markets. These examples show that wherever markets are created in close proximity to small producers, it helps in focusing buyers' attention as well as in reducing the information asymmetry regarding choices of customers. It also reduces the market-related search cost for SMEs. The challenge, then, is to link the local market to a distant consumer.

Policy makers who are facilitating establishment of clusters in the form of SEZs have a special lesson to learn from these examples. Organizing small producers to enhance their productivity and innovativeness, requires a network of complementary assets that overcome the disadvantage of size and access. This would require firms representing various elements of the supply chain (products as well as processes) to be co-located in such a cluster, so as to facilitate the formation of a network of product-developing and process-developing firms, as opposed to having firms of similar product type or similar processing capabilities being located together in a SEZ or a product park. It also opens up opportunities for increasing variety in such a network. It must also be recognized that the process of building capabilities is one that has to be facilitated and not left to chance. Consequently, the role of a coordinator must be emphasized. There appears to be a need for a market-driven coordinating agency (e.g., Gujarat Cooperative Milk Marketing Federation [GCMMF] in the AMUL network[51] or TAMA Association), which can earn the trust of the firms to coordinate the network on their behalf. There is an interesting model that is worth considering here—a network of diverse firms in terms of end products, intermediate products and processing capabilities, which may produce diverse products under the same umbrella brand of a marketing firm, which in turn, is equally owned by each of the production firms. The production firms own property rights on the marketing firm, and can exit without affecting the common brand. The marketing firm makes the quality and pricing decisions while the production firms make the production quantity and technology choices. The former also does the certification of the product quality in order to ensure a uniform offering from the network, and provides support for enhancing the capabilities of small producers. Variants of such models have been found to work effectively around the world.

Acknowledgement

The author wishes to thank Toshihiro Kodama of REITI, Japan, Kazunari Nagoi of TAMA, Japan, and Toshiro Wakayama of International University of Japan (IUJ), for helping him understand the details of the network enterprises in Japan as well as Liyan Zhang of Tianjin University, China for introducing him to Wenzhou. The author is also grateful to Rakesh Basant at Indian Institute of Management Ahmedabad (IIMA) for sharing many insights, over the years, on the cluster in Rajkot. This project was funded by the Thrust Area Grant from Ministry of Human Resource Development (MHRD), India.

Notes

1. P. Chandra and D. Tirupati, 'Business Strategies for Managing Complex Supply Chains in Large Emerging Economies: The Story of AMUL', *Working Paper*, Indian Institute of Management, Ahmedabad, 2003.
2. R. Jaikumar, 'Massimo Menichetti', *Case no. 686135*, Harvard Business School, Harvard University, Cambridge, 1986.
3. Based on field work in Japan during summer of 2004.
4. TAMA Document, 'Summary of TAMA', *mimeo*, Technology Advanced Metropolitan Area Association (TAMA), Tokyo, 2004.
5. T. Kodama, 'Industry-Academic and Inter-corporate Collaboration in TAMA', *RIETI Discussion Paper Series*, 02-E-004, Research Institute of Economy, Trade & Industry (RIETI), Tokyo, 2003; T. Kodama, 'Product Developing SMEs in TAMA', *PowerPoint Presentation Summary*, RIETI, Tokyo, 2004; T. Kodama, Personal Communication, 2004.
6. TAMA Document, 'Summary of TAMA'.
7. TAMA Association was founded by the efforts of Yuji Furukawa, Professor at the Tokyo University of Agriculture and Technology, Toshihiro Kodama of RIETI, and Makoto Ibuka of TAMA-TLO.
8. Kodama defines product-developing SMEs as SMEs that have product-designing capabilities, own their designs and products, and have a good understanding of the needs of the market. Product-processing SMEs are SMEs engaged in parts-processing (but do not have product development and marketing strengths), and whose strength lies in process technologies like cutting, grinding, sanding, casting, forging, pressing, coating, surface treatment, component assembly, and metal molding; Kodama, 'Product Developing SMEs in TAMA'.
9. Kodama, 'Product Developing SMEs in TAMA'.
10. 22 Yen = 1 USD in 2004.
11. TAMA Document, 'Summary of TAMA'.
12. SME Agency, 'SME Agency White Paper on SME 2000', 2002; T. Kodama, 'Product Developing SMEs in TAMA'.
13. 1 USD = 8.27 Renminbi Yuan (RMB) in 2002 (also called Yuan).
14. TDC Trade, 'Market Profiles on Chinese Cities and Provinces', Hong Kong Trade Development Council, January 2004, http://www.tdctrade.com/mktprof/china/wenzhou.htm
15. IFC, 'Scaling Up Private Sector Models for Poverty Reduction — A Report on the Field Visit to Sichuan and Zhejiang Provinces,

China', International Finance Corporation (World Bank Group), Washington DC, 2004.
16. K. Parris, 'Local Initiative and National Reform: The Wenzhou Model of Development', *The China Quarterly,* no. 134, 1993, 242–63.
17. Y-L Liu, 'Reforms from Below: The Private Economy and Local Politics in the Rural Industrialization of Wenzhou', The China Quarterly, 130, 1992, 293–316; A.P.L.Liu, 'The "Wenzhou Model" of Development and China's Modernization', *Asian Survey*, 32, 8, 1992, 696–711.
18. WMDPC, 'UN Industrial Development Organization Investigations of Cases in Wenzhou', Wenzhou Municipal Development and Planning Committee, Wenzhou, 2002.
19. Liu, 'Reforms from Below: The Private Economy and Local Politics in the Rural Industrialization of Wenzhou'.
20. J. Barber, 'Wenzhou: Capitalism Unbound', Globe and Mail, 23 October 2004.
21. WMDPC, 'UN Industrial Development Organization Investigations of Cases in Wenzhou'.
22. K. S. Tsai, 'Coping by Innovating', *mimeo*, Department of Political Science, Johns Hopkins University, Baltimore, MD; presented at the 2003 Annual Meeting of the American Political Science Association, Philadelphia, 27–31 August 2003.
23. S. Young, 'Policy, Practice and the Private Sector in China', The Australian Journal of Chinese Affairs, 21, 1989, 57–80; Liu, 'Reforms from Below: The Private Economy and Local Politics in the Rural Industrialization of Wenzhou'.
24. Liu, 'Reforms from Below: The Private Economy and Local Politics in the Rural Industrialization of Wenzhou'.
25. ADB Draft Report, 'The 2020 Project: Policy Support in the People's Republic of China', ADB Institute, 2001.
26. WMDPC, 'UN Industrial Development Organization Investigations of Cases in Wenzhou'.
27. Lin Bai, Jin Guo Wen, Zhou Yi Lin and Hu Fang Song, 'Markets in Wenzhou', Nanning: Guanxi Renmin Chubanshe, 1987, 99–101; reference cited in Liu, 'Reforms from Below: The Private Economy and Local Politics in the Rural Industrialization of Wenzhou'.
28. Tsai, 'Coping by Innovating'.
29. W. Wu, 'Entrepreneurs from Wenzhou: A Case Study of Economic Freedom in China,' *mimeo*, Department of Social &

Decision Sciences, Carnegie Mellon University and The Public Policy Programme, National University of Singapore, Singapore; paper presented at the 2004 Annual Meeting of the Public Choice Society, Baltimore, 11–14 March 2004.

30. Liu, 'The "Wenzhou Model" of Development and China's Modernization'
31. WMDPC, 'UN Industrial Development Organization Investigations of Cases in Wenzhou'.
32. Ibid.
33. Liu, 'Reforms from Below: The Private Economy and Local Politics in the Rural Industrialization of Wenzhou'.
34. T. Sonobe, D. Hu and K. Otsuka, 'From Inferior to Superior Products: An Inquiry into the Wenzhou Model of Industrial Development in China', Journal of Comparative Economies, 32, 3, 2004, 542–64.
35. Sonobe et al., 'From Inferior to Superior Products: An Inquiry into the Wenzhou Model of Industrial Development in China'.
36. Barber, 'Wenzhou: Capitalism Unbound'.
37. Sonobe et al., 'From Inferior to Superior Products: An Inquiry into the Wenzhou Model of Industrial Development in China'.
38. Ibid.
39. Ibid.
40. Ibid.
41. WMDPC, 'UN Industrial Development Organization Investigations of Cases in Wenzhou'.
42. This section is drawn from Basant (1997) and from author's fieldwork in Rajkot during 1999 and 2005.
43. USD 1 = Rupees (Rs.) 31.37 in 1995; USD 1= Rs 43.62 in 2005; 1 crore = 10 million.
44. Indian Express, 'Diesel Engine Industry Running out of Steam', Indian Express, 1 March 2005.
45. R. Basant, 'The Diesel Engine Industry Cluster Rajkot', Report, Indian Institute of Management, Ahmedabad, 1997.
46. Basant, 'The Diesel Engine Industry Cluster Rajkot'.
47. Ibid.
48. R. Basant and P. Chandra, 'Patel Brass Works', Case no. pro 242, Indian Institute of Management, Ahmedabad, 1997.
49. Indian Express, 'Rajkot Auto Parts Units are gearing up for Good Business', Indian Express, 11 May 2005.

50. R. Basant, P. Chandra and T. Sastry, 'Ancillarization of the Auto-component Sector in India: Strategies for Capability Building and Integration with Global Markets of Small Scale Firms', Report for Department of SSI, Government of India, Indian Institute of Management, Ahmedabad, 1999; R. Basant, P. Chandra and R. Upadhyayula, 'Building Technological Capabilities through Strategic Development of Industrial Clusters: A Study of the IT and Electronics Sectors in India', report under preparation, Indian Institute of Management, Ahmedabad, 2006; P. Chandra, 'Technology Characterization: Explaining a Few Things,' *mimeo*, Indian Institute of Management, Ahmedabad, 1995.
51. Chandra and Tirupati, 'Business Strategies for Managing Complex Supply Chains in Large Emerging Economies'.

Reference

R.B. Hayes and S.C. Wheelwright, 'Link Manufacturing Process and Product Life Cycle,' *Harvard Business Review*, January–February 1979, 2–9.

7

Insights in Research and Development, and Innovation in Europe and Spain in Particular: Practitioner Perspectives

V. Richard Benjamins, Lourdes Perez and Jesús Contreras

In this paper, we provide a practical insight into how innovation takes place in Europe, and with particular reference to Spain. Our objective is to contrast various types of information including (i) European Research and Development (R&D), innovation policy and activities, (ii) related R&D Key Performance Indicators (KPIs) as published by official institutions, and (iii) impact of innovation activities as measured by economic indicators. Our main observation is that there is a significant gap between European R&D investments made and the impact on the economic competitiveness of Europe. The underlying cause for these phenomena is mostly cultural, manifested by a mindset of European researchers who are more oriented towards research and less towards technology transfer and entrepreneurship. Changing such a mindset requires changes in the educational system, as well as in policy framework of European R&D and national R&D programs. The findings are based on 15 years of experience in R&D and innovation in the ICT sector, and hence cannot be generalized regarding other sectors.

Introduction

In 2000, the European Commission set itself an ambitious goal for the next decade to become the most competitive and dynamic knowledge-based economy in the world (i.e., by 2010). Now, ten years later, even though the stated goal is yet to be achieved, new ambitious objectives are set for 2020. In this paper, we will present a practitioner perspective—the activities and results of executing European R&D policies relating to Information and Communications Technology (ICT) across institutions, including universities, research organizations, small and medium-sized enterprises, start-ups and large enterprises.

The paper will examine the importance of the ICT sector using R&D expenditure in section 2; in section 3 of the paper, the authors briefly review the history and evolution of the Lisbon Council, and its outcome in terms of KPIs, and section 4 examines this phenomenon with reference to Spain. Then, in section 5, some questions and issues are formulated, that would explain key performance indicators using the authors' professional experience in innovation during the past 15 years. The paper concludes in section 6 with some guidelines for the future.

The ICT Sector – An Important Economic Contributor

In 2007, The ICT sector was responsible for 25 percent of the total business expenditure in R&D (BERD: Business Expenditure on Research and Development) in Europe, being by far the largest sector, followed by automotive (16 percent).[1] Figure 7.1 shows the distribution for the main sectors.

Figure 7.1 The ICT Sector Represents 25 percent of the Total Expenditure in R&D in European Industry

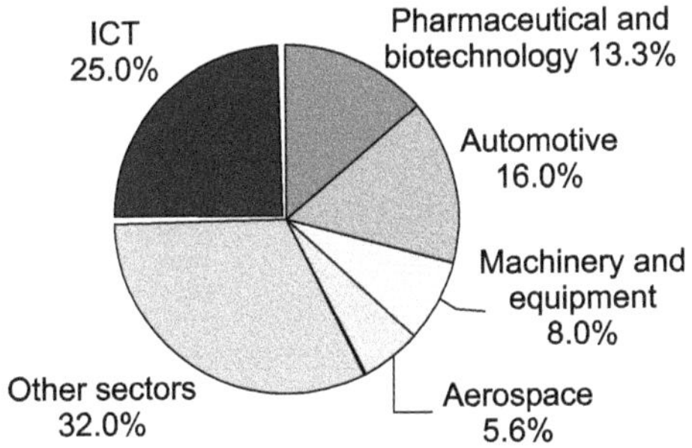

Source: The 2010 report on R&D in ICT in the European Union, JRC (Joint Research Centre), 2010.

Compared to the rest of the world, Europe's industry spends a larger portion of its R&D investment in non-ICT as can be seen in Figure 7.2.

Figure 7.2 Business R&D Spending on ICT versus Non-ICT in Different Regions of the World

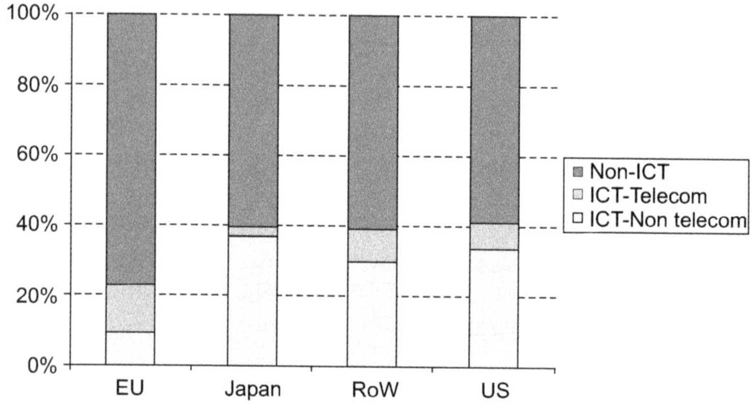

Source: The 2010 report on R&D in ICT in the European Union, JRC, 2010.

Note: ICT-Telecom includes Telecom Equipment and Telecom services.

The European Ambition and its Outcomes Ending 2010

After the European Commission set its ambitious goals during the Lisbon Council in 2000, the Barcelona European Council raised some issues, and noted that achieving the 2010 objectives were a real challenge:[2]

1. Many Europeans like to think of their continent as a global economic superpower. But when compared to the United States of America (USA) over the last decade or so, Europe looks like a laggard.
2. Even when European businesses do create new products, the dismal failure of the European Union (EU) to launch the community patent on schedule potentially jeopardizes intellectual property rights.

3. The process of starting a new business in the EU is slow and expensive. Once established, European businesses face a daunting burden of red tape.

Looking back at the past decade, we will review several KPIs of how Europe, and in particular Spain, is doing regarding R&D and Innovation. We review the main input parameters such as the number of researchers, publications, patents, gross expenditure on R&D (GERD), and venture capital.

1. Number of Researchers

An important indicator of the performance of R&D and innovation, may be availability of the number of researchers who could produce those results. If we compare Europe to the rest of the world, we can see that Europe has the second highest number of researchers in the world, and it has increased over the past decade (Figure 7.3).

Figure 7.3 Number of Researchers by World Region

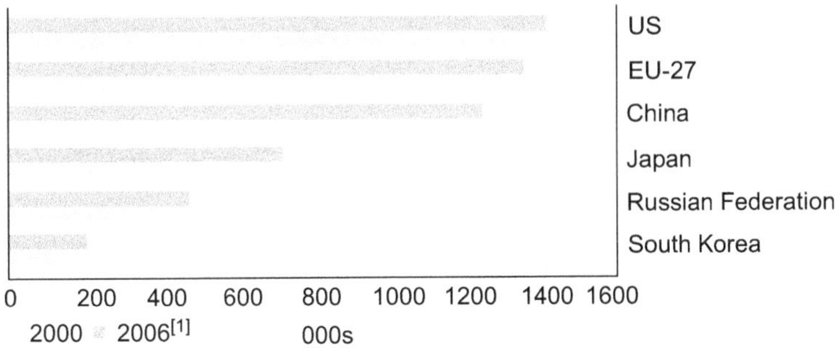

Source: DG Research in a more research-intensive and integrated European Research Area. Science, Technology and Competitiveness. Key figures report 2008/2009.

Data: Eurostat OECD

Note: [1] US: 2005

In absolute numbers as related to ICT only, the number of researchers in Europe has increased from about 190,000 in 2002 to 215,000 in 2007.[3] If we look at the number of researchers per European country, then we see that Germany, France and the United Kingdom (UK) form the top 3, with Spain at the fourth place with over 110,000 researchers (Figure 7.4).

2. Number of Publications

As expected, there seem to be a relationship between the number of researchers and number of publications, which is shown in Figure 7.5.

Figure 7.4 Number of Researchers per EU Country

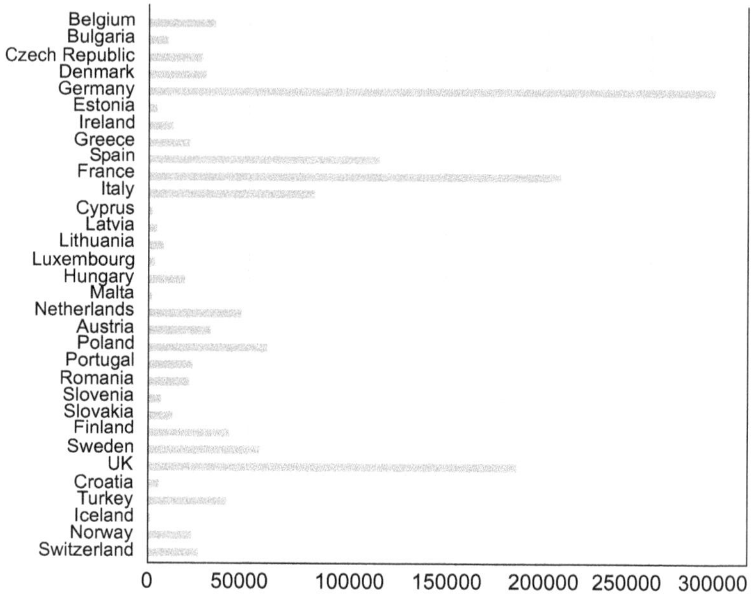

Source: DG Research in a more research-intensive and integrated European Research Area. Science, Technology and Competitiveness. Key figures report 2008/2009.

Data: Eurostal OECD

Note: [1] CH: 2004; FR, IT, PT, TR, IS, NO: 2005

Figure 7.5 Number of Publications per World Region

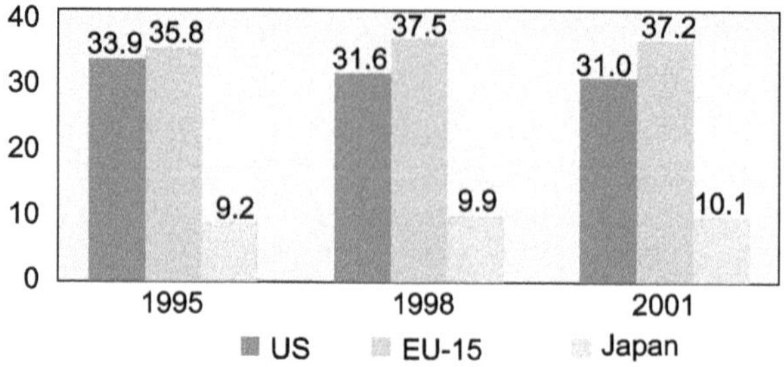

Source: Third European Report on S&T Indicators, 2003, 279.

3. Number of Patents

An examination of the number of patents indicate that, although the number of researchers in the USA and Europe are more or less the same, the number of patents is consistently higher in the USA (Figure 7.6). Regarding ICT, the USA patent activity is double that of Europe. If we compare the top ten European countries in patent applications with the USA, Germany is doing best and Spain, the worst (Figure 7.6). Although patents as such are no guarantee for transferring R&D results to the market and for economic impact, they are closer to the market than publications. Patents provide protection for inventions, and if companies want to commercialize patented inventions, they need to come to an agreement with the patent owner. Patents, thus, serve two purposes:

1. Generate an economic value (royalties or sale).
2. Serve as a measure to trace R&D results into the market and allow estimation of Return on Investment (ROI).

Figure 7.6 Patent Applications per Million Inhabitants in USA and Europe, Total and ICT.

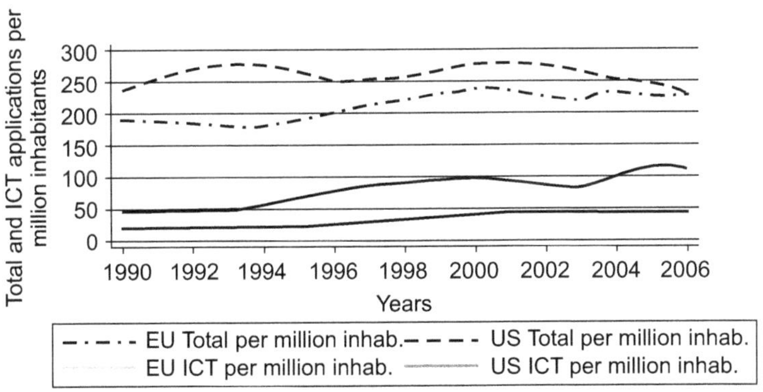

Source: JRC-IPTS calculations based on PATSTAT data (April 2009 release); Priority patent applications to the EPO, the 27 Member States National Patent Offices and the USPTO; Inventor criterion.

Source: The 2010 report on R&D in ICT in the European Union, 2010, 99.

Since Europe is doing worse on patents compared to the USA, it might indicate that Europe is less competitive than the USA. The same holds true within Europe: Spain, with the least amount of patents, is less competitive than other EU countries such as Germany and the Netherlands. Notice that in its current state, the patent industry is somewhat of a negotionting game. Many USA companies are patenting as much as possible in order to be able to negotiate in case of lawsuits for patent infringements. Moreover, in the USA—contrary to Europe—it is possible to patent software algorithms. So there are clear reasons why Europe is less active in patents than the USA. However, with the current rules of the game, more patents means more power, and it is hard to win the game without following the rules.

4. Gross Expenditure on R&D

The Government Expenditure on R & D (GERD) is an established measure of how much countries spend on R&D, usually separated in Government-financed and Business-financed or Business Expenditure on R & D (BERD). As can be clearly observed from Figure 7.8. Europe is significantly under spending compared to the USA and Japan.

Figure 7.7 Top Ten European Countries in Patents, Compared to the USA

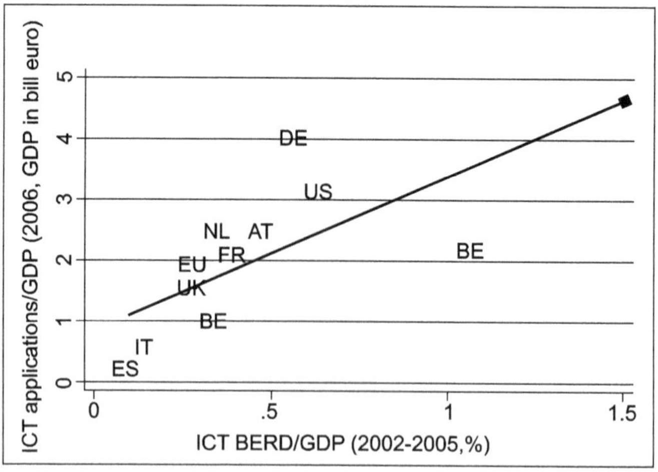

Source: IPTS calculation based on data from Eurostat, OECD, EU KLEMS and national statistics (million euro), and on PATSTAT data (April 2009 release). Priority patent applications to the EPO, the 27 Member States' National Patent Offices and the USPTO; Inventor criterion, JRC.

Figure 7.8 GERD Comparison of Europe against Other Regions

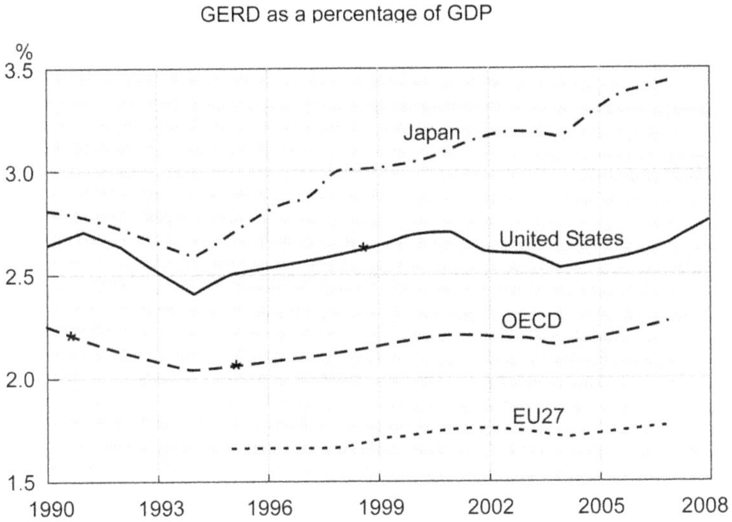

Source: Main Science and Technology 18 Indicators: Volume 2009(2), 18.

5. Venture Capital

Another usual indicator to measure the rate of technology transfer from R&D to the market is the use of Venture Capital. This measures two factors: (i) the availability of private money to invest in (risky and) innovative ideas to become successful in markets, and (ii) the cultural acceptance of entrepreneurship among researchers. If we compare Europe against the USA and Japan, we see that Europe is less active than the USA, though more active than Japan (Figure 7.9). We also see clearly the dot-com bubble and its explosion around the year 2000.

From an analysis of Europe based on, seed and start-up capital on the one hand, and expansion capital on the other, it could be noted that Finland has done very well, while Spain has done the least. As far as expansion is concerned, Spain remains as a most active country (Figure 7.10).

Figure 7.11 provides a comparison of Spain's innovation profile with that of the average Organisation for Economic Co-operation and Development (OECD) profile. As can be observed, Spain is doing well in terms of the number of researchers and publications, but is underperforming in terms of GERD and BERD, patents, and firms with new-to-market product innovations.

The authors believe that the Spanish culture is not the best one to create start-up companies, which have a high risk of failure. An anecdote evidencing this is that if parents are asked what job they prefer for their children, the preferred answer is 'civil servant' followed by a position at the Spanish incumbent telecommunications giant Telefonica. Asking the same question of students, results in the same answer. A civil servant in Spain has a job for life. And being an employee of Telefonica, in practice meant a synonym for a lifetime job (though this has changed in recent years). We see this as a manifestation of the fact that, in general, people in Spanish culture are willing to take only fewer risks than in other countries. Moreover, failure does not look good on one's curriculum vitae (CV). Another factor that serves as a barrier to risk-taking (e.g., setting up a start-up company) is that half of the Spanish labour market is very closed. It is expensive for employers to fire people as they are required to pay a significant amount of money to fired employees. This has become an obstacle to freely move the labour force, since people think twice before they leave a permanent job voluntarily, as they miss a lump sum payment in case of being fired. Moreover, employers perceive that a new hire will remain more economical for several years, as they will be the cheapest ones to get rid of in harder times. Thus, most people will avoid the risk of changing jobs, and are likely to stay on in their current jobs. This holds true for half of the labor market that is working on permanent contracts. The other half is based on temporary contracts, and they move around more easily as they have nothing to lose. If part of this population has a suitable background for starting technology-based companies, it will make a difference.

Figure 7.11 Spain's OECD Innovation Profile

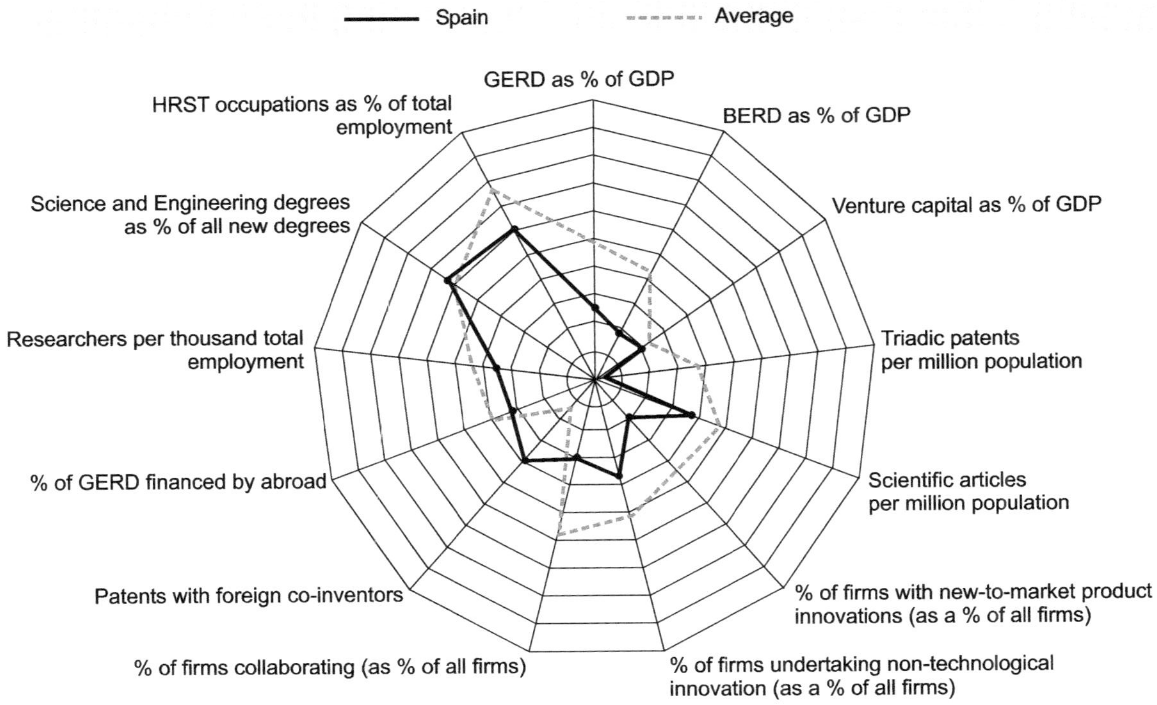

Source: OECD Science, Technology and Industry Outlook, 2008, 152

Interpretation of the Results

In the above section, we discussed a range of key performance indicators (KPIs) for R&D and innovation, and noted that Europe — and Spain, in particular — has not performed well as compared to the USA and Japan, while several input factors such as researcher and publications were comparable with the USA. So, what is 'wrong' with Europe? In this section, we will discuss some of the constraints in achieving its goals. The general assessment is based by authors on their collective experience in ICT industry for over 15 years in R&D, and innovation activities, and academic research.

As shown in Figure 7.12, only few ideas turn into successful products or services, and therefore innovation is a challenge. Coming up with new ideas, performing research, and product-development activities represent only the starting point. In addition, a commercial or social benefit needs to be associated with a new product or service, along with the right strategy to reach a target market. Thus, innovation requires not only R&D expertise but also marketing and sales capabilities.

From a practitioner's perspective, there are two major hurdles that need to be overcome before real innovation happens. The first one is related to the innovation funnel where many ideas enter and only few products or services find their way into the market.[4] At each funnel stage, a tollgate applies and a decision will be made whether to continue with the idea or to abandon it. The more an idea advances in the funnel, the more investment is needed to maintain that idea. That is, it is cheap to have ideas, but very expensive to develop and launch new products into the market.

The second hurdle is what Geoffrey Moore refers to as the 'chasm'.[5] Moore provides an excellent account of the technology life-cycle and how markets adopt a new technology over time. Early markets consist of innovators, real technical people who love technology for the sake of technology (the typical alpha and beta testers), and of visionaries, people who believe new technology can make a breakthrough in their business, and therefore are willing to take a certain risk. Mainstream markets are inhabited

Figure 7.12 Innovation is not Easy; only Few Ideas Turn into Successful Products or Services

Research across a range of industries shows that relatively few ideas turn into successful products or services

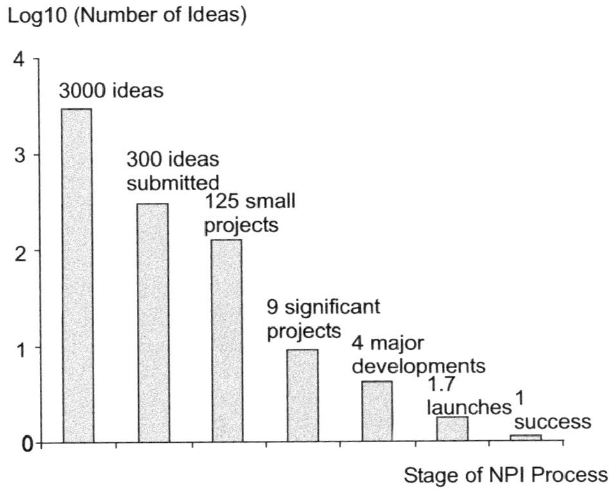

Innovation is difficult

- Despite the best efforts of remarkably talented people, most attempts to create commercially successful new products and services fail.
- Of every 3000 raw ideas only 1 ends up in a successful launch.
- ~60% of all-new product development efforst are discarded before they ever reach the market.
- Of the remaining 40%, 40% fail to become profitable and are withdrawn from the market.
- Consequence: 3/4 of all the money spent in product and services innovations results in products that do not succeed commercially (Source: *Wellsprings of knowledge*, D. Leonard, Harvard University Press, 1996).

Source: Dorothy Leonard-Barton, *Wellsprings of Knowledge: Building and Sustaining the Source of Innovation*, Harvard University Press, 1996.

by two types of customers: pragmatists, who believe that new technology can make a difference in their business, but are only willing to adopt the technology if there is sufficient evidence that the product works and does not cause additional risks. The conservative customers of the mainstream market will only buy new technology products if they think that not adopting them will cause problems in their business. Finally, the late market consists of laggards, who never buy technology unless it is embedded in devices they already know (Figure 7.13). Moore claims that there is a gap between the early adopters and the pragmatists, in the sense, that many high-tech products reach early adopters but never make the big step to the mainstream market, and thus fail to be profitable (and in our terminology 'fail to innovate'). In his book, Moore claims the percentage of failure to 'cross the chasm' to be as high as 90 percent.

Figure 7.13 The Technology Adoption Life-cycle as Defined by Geoffrey Moore (percentage)

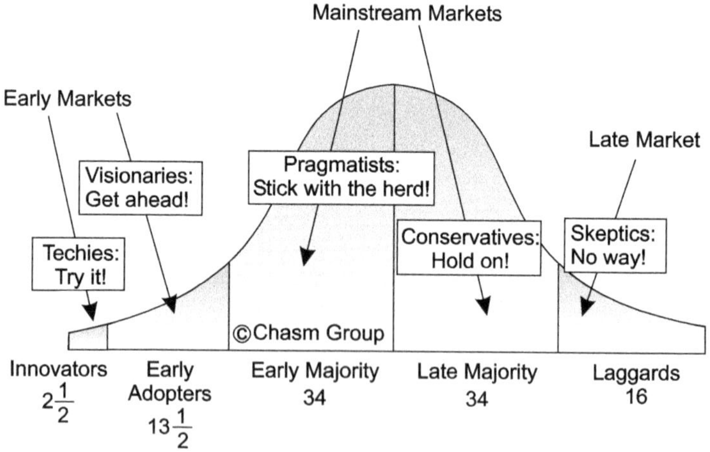

Source: Geoffrey Moore, *Crossing the Chasm: Marketing and Selling High-Tech Products to Mainstream Customers*, Harper Collins Publishers, 1999.

The authors allude that there is a difference between the USA and Europe, in the way actors in innovation deal with those two

Figure 7.14 Europe and USA Have Different Challenges for Bringing New Products to the Market

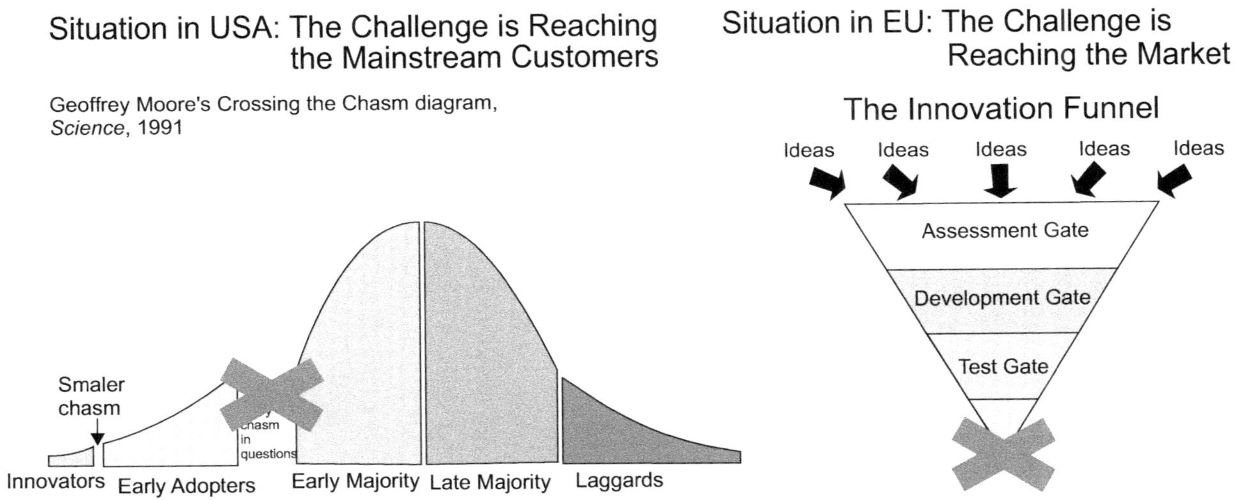

Source: Geoffrey Moore amd Regis McKenna, *Crossing the Chasm: Marketing and Selling High-Tech Products to Mainstream Customers*, Harper Collins Publishers, 1999.

hurdles, and this phenomena is presented in Figure 7.14.

While the challenge in the USA is to cross the chasm, in Europe the main problem occurs at an earlier stage, and that means only few results of R&D activities reach the market. This is a serious problem as the innovation funnel is broken in the sense that the last stage (transferring research results to the market) does not take place. The issue is, why does this occur? The authors have noted some factors that may explain this observation. As stated earlier, the generalizations made in the paper are solely based on authors' years of experience in performing R&D and innovation at various European organizations and companies. In addition, as competition has become more knowledge-based, academics have turned their attention to understanding how firms develop knowledge and the processes by which they convert knowledge into capabilities in order to meet market demands.

Cohen and Levinthal[6] originally defined absorptive capacity (ACAP) as the ability of a firm to recognize the value of new external knowledge, assimilate it and apply it to commercial ends as critical to its innovative capabilities and competitive advantage. Zahra and George[7] further distinguished between potential and realized absorptive capacity (PACAP and RACAP respectively). PACAP makes the firm receptive to acquiring and assimilating external knowledge but does not guarantee the exploitation of this knowledge. RACAP is a function of the transformation and exploitation capabilities of the firm, to leverage the knowledge that has been absorbed. PACAP and RACAP coexist at all times and have complementary roles. For example, firms cannot possibly exploit knowledge without first acquiring it. Similarly, firms can acquire and assimilate knowledge but might not have the capability to transform and exploit the knowledge for profit generation. This means that some firms possess a strong ingenuity to understand complex technical problems but are not as effective in translating such knowledge into product innovation strategies.

Firms acquire knowledge from different sources in their environment including inter-organizational relationships and business alliances. The diversity of these sources significantly influences the acquisition and assimilation capabilities. Past experience is closely related to new-product-development as experience reflects a firm's successes and failures over time. It can also significantly determine how firms acquire and assimilate new knowledge, as well as the locus of their future technological search. Thus, learning is a path-dependent process.

Certain events may act as triggers for firms to intensify their efforts to acquire new knowledge. For example, crises threatening a firm's existence or radical changes in the industry in which the firm operates, might serve as a trigger. Assimilation refers to the firm's processes that allow it to analyze, process, interpret, and understand the information obtained from external sources. It requires the sharing of relevant knowledge among members of a firm either through informal or formal mechanisms. Transformation denotes a firm's capability to develop and refine the ways, that facilitate combining existing knowledge and the newly acquired and assimilated knowledge. The expansion and complexity of knowledge have made it increasingly difficult for a single firm to contain and capitalize upon all relevant knowledge.[8] As a result, firms specialize and utilize inter-firm relationships and networks to complement and exploit their knowledge. Exploitation emphasizes the application of knowledge, and the ability to respond to market changes and opportunities.

Based on the European financial services firms, Jansen et al.[9] concluded that organizational mechanisms associated with coordination capabilities (cross-functional interfaces, participation, and job rotation) primarily enhance potential absorptive capacity, while organizational mechanisms associated with socialization capabilities primarily strengthen realized absorptive capacity. The limited empirical studies on ACAP seem to indicate that the lack of systematic routines and procedures to exploit knowledge along with the already mentioned socialization issues, contribute to the observed gap between European and the USA value-creation outcomes.

The USA and EU have different philosophies for R&D and Innovation

In the USA, funding bodies for research such as NSF (National Science Foundation) and DARPA (Defense Advanced Research Projects Agency) provide 100 percent funding of the research activities. The IPR (Intellectual Property Rights) is retained by the producing organization, but the government gets a free, eternal license to use any of the results. In Europe, (European Commission Framework Programs), the awards range between 50 percent (large enterprises) and 75 percent (SMEs, universities and research organizations) of funding, and the IPR is retained by the producing organizations; this is the situation in EU Framework Program 7(FP7). In the previous FPs, companies received 50 percent funding of all costs, and universities, 100 percent of marginal costs. The rationale for the EC to not provide 100 percent of funding is probably to avoid funding work that is not of any real interest to the organizations. It seems that the USA is willing to accept this risk. Another difference is that the USA research work is focused on research and its results, whereas in EU, research is a significant amount of effort and dedicated to justify spending of the taxpayers' money. That is, in the USA emphasis is on the 'use', while in Europe it is 'avoiding abuse'. A third difference lies in what is being considered a 'success' by researchers. In the USA, a successful research project results in market activity and thus in increased productivity and competitiveness, whereas in Europe success is defined by winning a new research project. This leads to more (continued) research activity, but not to improved productivity and competitiveness.

The best versus the right project

Another notable phenomenon of the past decade is that all funding proposals have become extremely competitive, especially in the IST (Information Society Technology) area (now called ICT again). In 2006, the acceptance rate of IST proposals was less than 15 percent. So, on an average, an organization had to submit about six projects for getting one approved. Obviously, this situation prompts applicants to try different projects, rather than to focus on the one best project. Due to multiple submissions, expert evaluators do their utmost to select the best projects. There is however a difference between the 'best' and the 'right' project. The best project guarantees the highest scientific excellence, and is formed by a consortium of organizations where each has a track record of excellent research. The 'right' project also takes into account other factors such as market and IPR opportunities, track record of previous successful innovations of the consortium members, etc.

The above-mentioned factors might provide some idea as to why Europe has so few global software players. Is the German company System Applications and Products (SAP) the only one? Europe does have researchers, it has excellent scientific publications, but it files fewer patents and has less venture capital. Is it then a cultural factor? Do EU researchers find research cool, while marketing is not done? Do EU researchers invent and publish, rather than protect IPR by filing patents? Is Europe more open-source-minded and patent-wary? Are EU researchers wary of the risks involved in starting technology-based companies? Is there an educational challenge? Should engineers, besides learning excellent engineering, also learn to value business aspects while at university? Should technical university curricula focus slightly more on what customers need rather than on technology alone? Several questions posed here do not have clear answers, but we hope that reflecting on them may help improve the economic and business impact of European ICT research.

The European Commission has adapted its instruments

The European Commission and the national governments have long been concerned about the issue of technology transfer of EC-funded R&D projects. Each project contract states that it is obligatory to include a work stream related to exploitation of the project results. However, in practice, this work stream usually gets less than 10 percent of the project resources, whereas according to the funnel theory, the further one moves in the funnel, the more investment is needed. Moreover, European researchers, in general, do not express much interest in exploitation work, and consider it more as a necessary evil to get the project funded. The result is what everybody knows: no follow up of the project and no activity beyond the project lifetime. European regulation on fair competition prohibits subsidizing commercial product development, and that is the reason why only limited exploitation efforts occur in research projects. However, in FP6 and continued in FP7, the Commission has created a new program called Competitive and Innovation Program (CIP) to fund projects that bring research results into the market place. Whereas in FP6, this had only 10 percent of funding (for complying with the regulation on fair competition), in FP7, CIP projects enjoy 50 percent of funding of the project costs. With time we will know whether this program improves technology transfer.

In conclusion, it may be stated that the strengths and weaknesses of European-funded and nationally-funded R&D projects are as follows:

Strengths

1. Create an infrastructure to perform research at the EU level
2. Create R&D jobs in research organizations and companies
3. Defining new R&D projects

Weaknesses

1. No systematic technology transfer from research to the market
2. Low number of ICT product start-ups
3. Low contribution to competitiveness of EU in ICT product business

Europe is not giving up: The ambitious objectives for 2020

Europe keeps hoping to improve its competitiveness, and in getting better results from research and innovation programs. For 2020, it has formulated several objectives, plans and programs for the EU to become a smart, sustainable and inclusive economy.[10] The Commission is expected to put forward seven flagship initiatives to make progress under each of the seven priority themes; including the creation of an 'Innovation Union' to improve framework conditions and access to finance for research and innovation; and ensure that innovative ideas can be turned into products and services that create growth and jobs, create 'an industrial policy for the globalization era' that would improve the business environment, notably for SMEs, and support the development of a strong and sustainable industrial base to compete globally.

Whether in 2020, the European economy will be 'the most competitive and dynamic knowledge-based economy in the world' will be known with time. But the authors believe that European research in ICT has to undergo a change that should start from university education, and cannot be solved only by pumping in more money into current research institutions.

Directions for the Future

While recognizing that the European situation requires long-term changes, the paper presents some guidelines to improve the management of R&D and innovation. The guidelines aim to 'restore' the broken innovation funnel, and this idea is drawn from the changes introduced in Telefonica R&D, one of Europe's largest R&D companies, and Spain's largest. The underlying rationale of the guidelines is to take the correct decisions at important tollgates, that would help move product ideas through the innovation funnel from research to the market. We have to remember that the more a product progresses through the funnel, the more investments are required to continue working on it. Wrong decisions will result in higher costs. The guidelines include:

1. **Technological difference:** It is important that the technology developed be differential. Otherwise, there will be plenty of alternative technologies, either already available or easy to replicate.

2. **Endorsed innovation:** As said, moving through the funnel requires increasingly deeper investment pockets. Finding a market player that confirms interest in the technology or product throughout the funnel, helps in making the correct bet.
3. **Market competition:** It is good to force comparisons of the product under development with existing alternatives in the market. Engineers tend to focus on technological features some of which already may be available in existing products. Making a product under development compete with existing products in the market might help avoid investing resources in problems that either do not exist or are already solved.
4. **Partner for industrialization:** Do not always try to develop the commercial product completely by yourself. It may be more interesting to focus—as a research institute—on the core technology, and leave the industrialization of the product to a partner. IPR for the core technology can thus be retained and evolved, while the more market-oriented activities are taken care of by the partner.
5. **Licensing:** Following a licensing approach is good for several reasons. First, it requires defining an IPR policy, e.g., through patenting, which is—as we have seen in this article—not a standard procedure for European research. Second, it forces the product to be competitive, i.e. clients may not be interested in paying a license fee for something that has no positive business impact. Third, it facilitates tracing the results of the product (innovation) in the market by keeping track of economic transactions.

While those guidelines do not guarantee successful innovation—indeed, by definition many innovations fail and never reach the market—they help focusing on the right factors when taking decisions for continuing to work on the product/idea. It goes without saying that the guidelines will work better in research environments with an entrepreneurial spirit, than in settings which are currently typical of European research.

Notes

1. Throughout this article we use many figures from public reports produced by the European Commission with KPIs on innovation. We use the figures as they were published originally, clearly indicating the source. We keep the original 'meta-data' of the figures for easy reference to the original publication.
2. E. Bannerman, 'The Barcelona European Council', March 2002, http://www.euractiv.com/en/innovation/barcelona-european-council/article-110191 (accessed 21 August 2013).
3. Geomina Turlea, Daniel Nepelski, Giuditta de Prato, Sven Lindmark, Andrea de Panizza, Lucio Picci, Paul Desruelle and David Broster, 'The 2010, report on R&D in ICT in the European Union', 2010, http://ftp.jrc.es/EURdoc/JRC57808.pdf. (accessed 16 November 2012)
4. H. Chesbrough, 'Open Innovation: The New Imperative for Creating and Profiting from Technology', Harvard Business Press, 2003.
5. Geoffrey Moore, *Crossing the Chasm: Marketing and Selling High-tech Products to Mainstream Customers*, Harper Collins Publishers, 1999.
6. W. M. Cohen, D. A. Levinthal 'Absorptive Capacity: A New Perspective on Learning and Innovation', *Administrative Science Quarterly*, 35, 1990, 128–52.
7. George Zahra, 'Absorptive Capacity: A Review, Reconceptualization and Extension', *Academy of Management Review*, 27(2), 2002, 185–203.
8. P. J. Lane, B. R. Koka and S. Pathak, 'The Reification of Absorptive Capacity: A Critical Review and Rejuvenation of the Construct', *Academy of Management Review*, 31(4), 2006, 833–63; A. G. Pateli, 'Decision Making on Governance of Strategic Technology Alliances', *Management Decision*, 47(2), 2009, 246–70.
9. Frans A.J. van den Bosch, Justin J. P. Jansen and Henk W. Volberda, 'Managing Potential and Realized Absorptive Capacity: How Do Organizational Antecedents Matter?', *Academy of Management Journal*, 48(6), 2005, 999–1015.
10. European Commission, 'A Strategy for Smart, Sustainable and Inclusive Growth, 2010', 2010, http://ec.europa.eu/europe2020/index_en.htm (accessed 21 August 2013).

8

Building Innovative Teams on Diverse Creativity

Vesa Routamaa

To be competitive, companies need innovations, for example, in terms of organization, technology, products, and marketing. Innovations require new ideas, and new ideas require creativity. Innovation is the follow-through on creation. Creativity is thinking up new ideas and innovation is doing new things. Creativity produces seeds for innovation. However, creativity is not only new ideas. To be able to grow the seed for innovation also needs daring. That is, creativity is a combination of creation and daring. Creativity can be manifold, and different people can have different creativity. Some personality types may bring out more and more daringly new ideas whereas some types bottle them up. That is, it is important which kind of individual or team is responsible for creating ideas for innovations. In spite of a remarkable lack of research on innovative teams, it has been found that effective teams are heterogeneous. In practice, unfortunately, a majority of teams are built without any method to get together reasonably heterogeneous, creative teams. Added value does not grow up putting similar people and similar strengths together. In this study, personality differences are analyzed in relation to creativity-orientation of individuals in order to find out combinations of personality types to design creative and innovative teams. The Myers–Briggs Type Indicator® was used in the study to measure the personality types. A four-dimension model of creativity orientation was obtained using factor analysis to indicate creativity orientations. Several significant statistical relationships were found. Typical representatives of each creativity dimension will be reported.

Innovativeness is competitive advantage to any company. Innovativeness requires appropriate people and organization. In western culture, individual creativity and big inventions have been favored whereas in eastern culture, gradual building of innovations by collective work has been more typical. Both ways of innovation development need creativity and creative people. It is not irrelevant which kind of individual or team is responsible for creating ideas for innovations. Further, an individual always has a very limited cognitive view, and also, teams should be built on different personalities to achieve all-around view on problems.[1] In practice, unfortunately, a majority of teams are built without any method to form reasonably heterogeneous, creative teams. Putting similar people and similar strengths together does not add value to the team. Creativity can be manifold, and different people can have different types of creativity. Some personality types may daringly bring out new ideas, whereas some types bottle them up. Most people have some kind of creativity, but personality types separate individuals in terms of creation and daring.

To be able to find creative people for different purposes or to build creative teams, creative orientations of different personalities should be known. Most people have some kind of creativity or certain role in creative groups or teams.[2] That is why the relationship between personality types and creativity orientations must be studied. The basic question of this study is whether there is to be found an association between personality types and creative behavior in terms of creation and daring. This research question requires appropriate typology of personalities to be able to accomplish research results, which can be applied in practical building or creative teams at different phases of innovation process from ideas to realized innovation. For this purpose, Jung's typology and the Myers–Briggs Type Indicator (MBTI) based on it was applied. The description of characteristics and behaviors of different personality types suggest that types differ in terms of creativity and daring, but more empirical evidence was needed. There is also lack of relevant creativity measurements. In this study, first, a unique four-dimension model of creativity behavior is empirically developed. Thereafter, personality preferences and types are analyzed in relation to creativity orientation.

Personality – Beginnings of Creativity and Innovation

Personalities differ in terms of creation and daring. There are several tests to measure personality, but only few to conceptualize personality based on a theory to illustrate creativity and daring behavior. In this study, the Myers–Briggs Type Indicator was used. It is based on Carl Jung's theory of psychological types, and it reports personality preferences on four scales: Jungian Extroversion — Introversion, Sensing — Intuition, Thinking—Feeling, and the Judging—Perceiving preference added by Briggs and Myers.[3] According to Myers:[4]

> [T]he MBTI is primarily concerned with the valuable differences in people that result from where they like to focus their attention, the way they like to take information, the way they like to decide, and the way they like to adopt.

Usually one pole dominates over another. The eight preferences are identified in 16 types, each representing a certain preference order.[5] Briefly illustrated, the preferences or dimensions are:[6]

Extroversion (E)	Interested in people and things in the world around them. Breadth of interests.
Introversion (I)	Interested in the ideas in their minds that explain the world. Depth of concentration.
Sensing (S)	Interested in what is real and can be seen, heard and touched. Reliance on facts.
Intuition (N)	Interested in what can be imagined and seen with 'the mind's eye'. Grasp of possibilities.
Thinking (T)	Interested in what is logical and works by cause and effect.
Feeling (F)	Interested in knowing what is important and valuable. Warmth and sympathy.

Judging (J)	Interested in acting by organizing, planning and deciding. Organization.
Perceiving (P)	Interested in acting by watching, trying out and adapting. Adaptability.

As stated by Myers and McCaulley:[7]

> According to theory, each of the 16 types results from a preference for one pole of each of the four preferences over the opposite pole. A preference of any dimension is designed to be psychometrically independent of the preferences of the other three dichotomies, so that the preferences on the four dichotomies yield 16 possible combinations called types, denoted by the four letters identifying the poles preferred (e.g., ESTJ, INFP). The theory postulates specific dynamic relationships between the preferences. For each type, one process (S/N or T/F) is the leading or dominant process and a second process serves as an auxiliary. Each type has its own pattern of dominant and auxiliary processes and the attitudes (E or I) in which these are habitually used. Determining these dynamic relationships is enabled by the J-P dichotomy of the MBTI. The characteristics of each type follow from the dynamic interplay of these processes and attitudes.

In order to interpret the association between type and creative behavior, the types are next briefly illustrated:[8]

ISTJs: Quiet and serious. Succeed through concentration and thoroughness. Practical, orderly, matter-of-fact, logical, realistic, and dependable. See to it that everything is well-organized. Take responsibility. Make up their own minds as to what should be accomplished and work toward it steadily, regardless of protests or distractions.

ISFJs: Quiet, friendly, responsible, and conscientious. Work devotedly to meet their obligations. Lend stability to any project or group. Thorough, painstaking, accurate. Their interests are usually not technical. Can be patient with necessary details. Loyal, considerate, perceptive, concerned with how other people feel.

INFJs: Succeed by perseverance, originality, and desire to do whatever is needed or wanted. Put their best efforts into their work. Quietly forceful, conscientious, concerned for others. Respected for their firm principles. Likely to be honored and followed for their clear visions as to how best to serve the common good.

INTJs: Have original minds and great drive for their own ideas and purposes. Have long-range vision and quickly find meaningful patterns in external events. In fields that appeal to them, they have a fine power to organize a job and carry it through. Skeptical, critical, independent, determined. Have high standards of competence and performance.

ISTPs: Cool onlookers, quiet, reserved, observing and analyzing life with detached curiosity and unexpected flashes of original humor. Usually interested in cause and effect, how and why mechanical things work, and in organizing facts using logical principles. Excellent at getting to the core of a practical problem and finding the solution.

ISFPs: Retiring, quietly friendly, sensitive, kind, and modest about their abilities. Shun disagreements; do not force their opinions or values on others. Usually do not care to lead but are often loyal followers. Often relaxed about getting things done because they enjoy the present moment and do not want to spoil it by undue haste or exertion.

INFPs: Quiet observers, idealistic, loyal. Important that outer life be congruent with inner values. Curious, quick to see possibilities, often serve as catalysts to implement ideas. Adaptable, flexible and accepting unless a value is threatened. Want to understand people and ways of fulfilling human potential. Little concern with possessions or surroundings.

INTPs: Quiet and reserved. Especially enjoy theoretical or scientific pursuits. Like solving problems with logic and analysis. Interested mainly in ideas, with little liking for parties or small talk. Tend to have sharply defined interests. Need careers where some strong interest can be used and useful.

ESTPs: Good at on-the-spot problem-solving. Like action, enjoy whatever comes along. Tend to like mechanical things and sports, with friends on the side. Adaptable, tolerant, pragmatic, focused on getting results. Dislike long explanations. Are best with real things that can be worked, handled, taken apart, or put together.

ESFPs: Outgoing, accepting, friendly, enjoy everything and make things more fun for others by their enjoyment. Like action and making things happen. Know what's going on and join in eagerly. Find remembering facts easier than mastering theories. Are best in situations that need sound common sense and practical ability with people.

ENFPs: Warmly enthusiastic, high-spirited, ingenious, imaginative. Able to do almost anything that interests them. Quick with a solution to any difficulty and ready to help anyone with a problem. Often rely on their ability to improvize instead of preparing in advance. Can usually find compelling reasons for whatever they want.

ENTPs: Quick, ingenious, good at many things. Stimulating company, alert and outspoken. May argue for fun on either side of a question. Resourceful in solving new and challenging problems, but may neglect routine assignments. Apt to turn to one new interest after another. Skillful in finding logical reasons for what they want.

ESTJs: Practical, realistic, matter-of-fact, with a natural head for business or mechanics. Not interested in abstract theories, want learning to have direct and immediate application. Like to organize and run activities. Often make good administrators; are decisive, quickly move to implement decisions; take care of routine details.

ESFJs: Warm-hearted, talkative, popular, conscientious, born co-operators, active committee members. Need harmony and may be good at creating it. Always doing something nice for someone. Work best with encouragement and praise. Main interest is in things that directly and visibly affect people's lives.

ENFJs: Responsive and responsible. Feel real concern for what others think or want, and try to handle things with due regard for others' feelings. Can present a proposal or lead a group discussion with ease and tact. Sociable, popular, sympathetic. Responsive to praise and criticism. Like to facilitate others, and enable people to achieve their potential.

ENTJs: Frank, decisive, leaders in activities. Develop and implement comprehensive systems to solve organizational problems. Good at anything that requires reasoning and intelligent talk, such as public speaking. Are usually well informed and enjoy adding to their fund of knowledge.

Instead of the 16 types, different types of cognition have also been explained with the help of four temperaments,[9] which also give hints of creativity behavior of the types. The temperaments are **SP**, sensation perceptive **Artisan** (amiable, troubleshooter); **SJ**, sensation judging **Guardian** (driver, traditionalist); **NF**, intuitive feeling **Idealist** (catalyst, expressive) and **NT**, intuitive thinking **Rational** (visionary, analytical). SP has often been mentioned as the typical entrepreneur — negotiates well, is good in a crisis and is a risk-taker. NT is a visionary who enjoys complexity and is an architect of change, sees long- and short-term implications, and focuses on possibilities.[10] Carland, Carland and Higgs[11] found NTs (intuitive thinking) displaying the highest entrepreneurship tendency, i.e., NTs, as distinguished from the other temperaments, fit the traditional view of entrepreneurship in that the NT preference was highly correlated with innovation.[12] These results uncovered that entrepreneurs tend to be NTs but also NPs. Entrepreneurs are found to be innovative and risk-taking. Routamaa[13] reported that most typical entrepreneurial types are ESFP, ESTP, INTP, ISTP, ENTP, and ENFP.

According to Myers and McCaulley,[14] creativity is expected to be associated primarily with the preference for intuition (N), and secondarily, with the preference for perception (P). Later studies have supported the association between intuition and perception and creativity and originality. Measured by the Kirton Adaption Innovation Inventory, Gryskiewicz and Tullar,[15] Van Rooyen,[16] and Carne and Kirton[17] found that the preferences for N and P were associated with an innovative problem-style. Carne and Kirton[18] suggested that these results indicate one's style of creativity instead of the level of creativity. According to McCaulley,[19] riskier decisions might be expected from extroverted (E), intuitive (N), and feeling (F) types.

Asikainen and Routamaa[20] studied the relationship between the MBTI and creativity orientations of managers. The creativity orientations based on social risk-taking and creativity were organizational performers, pragmatic challengers, resourceful thinkers and innovators. The nature of those orientations can be deduced from Figure 8.1. Asikainen and Routamaa's[21] (1997) results corresponded quite a lot with McCaulley's[22] view.

Figure 8.1 The Relationship between MBTI and Creativity Orientations[23]

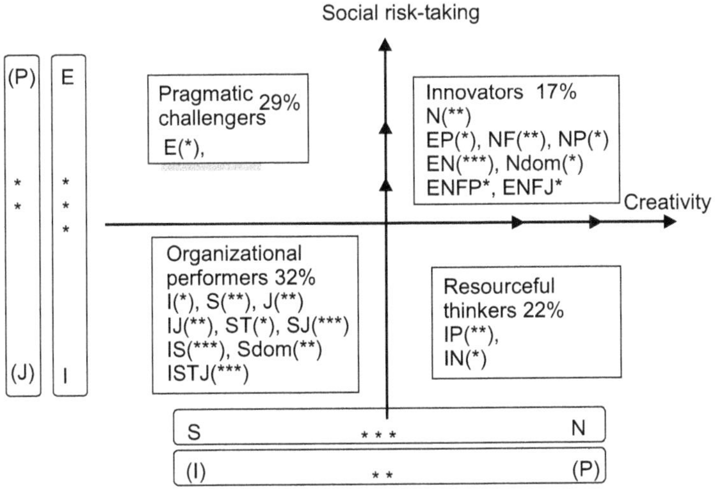

Source: V. Asikainen and V. Routamaa, *The Relationship between the MBTI and the Creativity Orientations of Managers*, Proceedings of the Second Biennial International Conference on Leadership of the Center for Applications of Psychological Type, Gainesville, April 1997, 11–16.

Based on the studies on entrepreneurial and creativity studies, there is evident association between personality types and creativity behavior in terms of creation and risk-taking.

Method

A Finnish version of the Myers–Briggs Type Indicator, validated by the author, was used in the study to measure the personality types. The construct validity and reliability of the indicator has been presented, for example, by Asikainen.[24] Measuring creativity was very challenging. At best it could be some kind of laboratory testing but, in practice, it was not possible in correlation research. In general, the measures of creativity or creation are non-validated questionnaires indicating attitudes, opinions, manners, etc. For example, Asikainen and Routamaa[25] had to change Byrd's[26] Creation and Risk-Taking (C&RT) measure radically, because, at least in Finnish culture, even the items of two dimensions, creation and risk-taking, measured wrong dimensions. A totally new measure was constructed. Analyzing 332 observations answering some 230 items, a unique four-dimension model of creativity orientation was obtained using factor analysis to indicate creativity orientations. Four factors, i.e., the dimensions of creativity behavior obtained were: (1) Meticulous planner, (2) Individualistic thinker, (3) Creative idea man, and (4) Creative law challenger. Internal consistencies of all four dimensions were at least satisfactory, with all Cronbach alphas over 75.

The dimensions of creativity behavior, creativity orientations are described as follows:

Meticulous planner

(i) Lives present, do not leave for tomorrow what can be done today.
(ii) Dutiful, systematic, do not stand disorganization.
(iii) Punctual.
(iv) Favor logical and well-defined methods.
(v) First work, then amusement.
(vi) Laborious. Work is the key for success.
(vii) Sees himself/herself trustful and responsible.

(viii) Do not like insecurity or surprises.

Individualistic thinker

(i) Lives as one likes to live one's own way, wants to be himself/herself without caring for others' opinions.
(ii) Believes in being street-smart.
(iii) Takes risk to express his/her opinion without caring what others think.
(iv) Social, not enjoying alone.
(v) Assertive, accepts own vices.

Creative idea man

(i) Not necessarily feet on the ground.
(ii) Values creative people and imagination.
(iii) Questions things even without solutions.
(iv) Believes own uniqueness and dreams are important sources of innovations.
(v) Thinks new ideas without taking care to see if they are useful or not.
(vi) Trusts own intuition in problem-solving, and questions profundities.

Creative law challenger

(i) Self-reliant. Trusts that own ideas are best.
(ii) Difficult person even in his/her own eyes.
(iii) May break moral issues, wave aside the rules.
(iv) May be temperamental having strong feelings.
(v) Gets bored easily, and feels sometimes other people boring and uninteresting.
(vi) Independent; does not want to be led.
(vii) Values self-confident people, and may consider himself/herself to be better than others.

In all, 189 respondents filled both the MBTI form and creativity orientation form for this study. Using SRTT (The Selection Ratio Type Table),[27] 52 meticulous planners, 54 individualistic thinkers, 44 creative idea men, and 53 law challengers of highest quartiles in each dimension were compared to the whole sample.

Table 8.1 Type Distribution of the Personality Types in the Total Sample of the Study

ISTJ N = 14 % = 7.41	**ISFJ** N = 6 % = 3.17	**INFJ** N = 1 % = 0.53	**INTJ** N = 4 % = 2.12
ISTP N = 10 % = 5.29	**ISFP** N = 4 % = 2.12	**INFP** N = 1 % = 0.53	**INTP** N = 5 % = 2.65
ESTP N = 12 % = 6.35	**ESFP** N = 10 % = 5.29	**ENFP** N = 21 % = 11.11	**ENTP** N = 16 % = 8.47
ESTJ N = 37 % = 19.58	**ESFJ** N = 20 % = 10.58	**ENFJ** N = 5 % = 2.85	**ENTJ** N = 23 % = 12.17

N = 189

Results

Next, the relationship between personality and creativity orientations will be presented. In the case of statistical significant relationships, the significance is marked in connection to the abbreviations (*=.05; **=.01; ***=.0001).

Meticulous Planners

Sensing (S*) and judging (J**) preferences are typical for meticulous planners. Also pairs introvert-judging (IJ*), extrovert-sensing (ES**), thinking-judging (TJ**), and temperament sensing-judging (SJ**).

At the type level, ESTJs*, and, even though not significantly, also ISTJs, INTJs and ESFJs are overrepresented among meticulous planners. Instead, intuitive (Ns*), spontaneous (Ps**), extrovert-spontaneous (EP*), intuitive-spontaneous (NPs**), thinking-spontaneous (TP**), and extrovert-intuitions (EN**) are underrepresented among meticulous planners. Also Ndoms* are not common in this group of orientation. Concerning types, all the P-types except ESFP are underrepresented among meticulous planners even though not significantly, because of the small number of observations (Table 8.2).

Individualistic Thinkers

Extroverts (Es**) only tend to be statistically significantly individualistic thinkers. At the type level, there are no statistically significant relationships but ESTJs are overrepresented in this group.

Introverts (Is*) and introvert-sensings (ISs*) are significantly underrepresented among individualistic thinkers. Sdoms also are few in this group of orientation even though not significantly. Similarly all introverted types excluding INFP. It may be noted that individualistic thinking means not solitary thinking which may be typical for many introverts (Table 8.3).

Table 8.2 Meticulous Planners Compared to the Total Sample

Relationship Between Type and Creativity		**Meticulous Planner** compared with Whole Base N = 52	
ISTJ N = 14 % = 15.38 I = 2.08 CCCCCCCC	**ISFJ** N = 2 % = 3.85 I = 1.21 CC	**INFJ** N = 0 % = 0.00 I = 0.00	**ISTJ** N = 3 % = 5.77 I = 2.73 CCC
ISTP N = 0 % = 0.00 1 = 0.00	**ISFP** N = 1 % = 1.92 I = 0.91 C	**INFP** N = 0 % = 0.00 1 = 0.00	**INTP** N = 0 % = 0.00 1 = 0.00
ESTP N = 2 % = 3.85 I = 0.61 CC	**ESFP** N = 3 % = 5.77 I = 1.09 CCC	**ENFP** N = 1 % = 1.92 I = 0.17 C	**ENTP** N = 1 % = 1.93 I = 0.23 C
ESTJ N = 17 % = 32.69 I = 1.67* CCCCCCCCCC CCCCCCC	**ESFJ** N = 8 % = 15.38 I = 1.45 CCCCCCCC	**ENFJ** N = 1 % = 1.92 I = 0.73 C	**ENTJ** N = 5 % = 9.62 I = 0.79 CCCCC

Print date:0

	N	%	I
E	38	73.08	0.96
I	14	26.92	1.13
S	41	78.85	1.32*
N	11	21.15	0.53*
T	36	69.23	1.08
F	6	30.77	0.86
J	44	84.62	1.45***
P	8	15.38	0.37***
IJ	13	25.00	1.89*
IP	1	1.92	0.18*
EP	7	13.46	0.43*
EJ	31	59.62	1.33
ST	27	51.92	1.34
SF	14	26.92	1.27
NF	2	3.85	0.26
NT	9	17.31	0.68
SJ	35	67.31	1.65***
SP	6	11.54	0.61
NP	2	3.85	0.17**
NJ	9	17.31	0.99
TJ	33	63.46	1.54**
TP	3	5.77	0.25**
FP	5	9.62	0.50
FJ	11	21.15	1.25
IN	3	5.77	0.99
EN	8	15.38	0.45**
IS	11	21.15	1.18
ES	30	57.69	1.38**
Sdom	15	28.85	1.30
Ndom	5	9.62	0.43*
Tdom	22	42.31	1.07
Fdom	10	19.23	1.21

* < .05, ** <.01, *** < .001

Source: Empirical Studies based on The Selection Ratio Type Table, see Granade and Myers, 1987
Note: C = 1 person.
Base total N = 189. Groups are independent.

Tables 2–5: I = 1, the proportion of the variable is identical to whole sample; I > 1, bigger; I < 1, smaller.

Table 8.3 Individualistic Thinkers Compared to the Total Sample

Relationship Between Type and Creativity

Individualistic Thinker
compared with
Whole Base
N = 54

ISTJ	ISFJ	INFJ	INTJ
N = 2 % = 3.70 I = 0.50 CC	N = 0 % = 0.00 I = 0.00	N = 0 % = 0.00 I = 0.00	N = 0 % = 0.00 I = 0.00
ISTP N = 0 % = 0.00 I = 0.00	**ISFP** N = 1 % = 1.85 I = 0.88 C	**INFP** N = 1 % = 1.85 I = 3.50 C	**INTP** N = 0 % = 0.00 I = 0.00
ESTP N = 4 % = 7.41 I = 1.17 CCCC	**ESFP** N = 1 % = 1.85 I = 0.35 C	**ENFP** N = 7 % = 12.96 I = 1.17 CCCCCCC	**ENTP** N = 6 % = 11.11 I = 1.31 CCCCCC
ESTJ N = 17 % = 31.46 I = 1.61 CCCCCCCCCC CCCCCCC	**ESFJ** N = 4 % = 7.41 I = 0.70 CCCC	**ENFJ** N = 2 % = 3.70 I = 1.40 CC	**ENTJ** N = 9 % = 16.67 I = 1.37 CCCCCCCCC

	N	%	I
E	50	92.59	1.22**
I	4	7.41	0.31*
S	29	53.70	0.90
N	25	46.30	1.15
T	38	70.37	1.10
F	16	29.63	0.82
J	34	62.96	1.08
P	20	37.04	0.89
IJ	2	3.70	0.28
IP	2	3.70	0.35
EP	18	33.33	1.07
EJ	32	59.26	1.32
ST	23	42.59	1.10
SF	6	11.11	0.53
NF	10	18.52	1.25
NT	15	27.78	1.09
SJ	23	42.59	1.05
SP	6	11.11	0.58
NP	14	25.93	1.14
NJ	11	20.37	1.17
TJ	28	51.85	1.26
TP	10	18.52	0.81
FP	10	18.52	0.97
FJ	6	11.11	0.66
IN	1	1.85	0.32
EN	24	44.44	1.29
IS	3	5.56	0.31*
ES	26	48.15	1.15
Sdom	7	12.96	0.58
Ndom	13	24.07	1.08
Tdom	26	48.15	1.21
Fdom	8	14.81	0.93

Print date:0

* < .05, ** < .01, *** < .001

Source: Empirical Studies based on The Selection Ratio Type Table, see Granade and Myers, 1987

Note: C = 1person

Base total N = 189. Groups are independent.

Creative Idea Men

Intuitives (Ns***), intuitive-feelers (NFs**), intuitive-spontaneous (NPs**), and extrovert-intuitives (ENs***) and correspondingly N**doms tend to produce new ideas. At the type level, ENFPs** are statistically overrepresented but also, for example, ENTPs, ENTJs, are also overrepresented in this group.

Sensing (S**), introvert-judging (IJ*), sensing-thinking (ST*), sensing-judging (SJ**), extrovert-sensing (ES**) and Sdoms** do not tend to pump up new ideas. Especially, ISTJs are totally missing from this group, and also, for example ISFJs, ESTJs, ESTPs, ESFPs and ESFJs are very clearly underrepresented (Table 8.4).

Again, the division of labor of preferences and types were very clear in the roles of creativity orientation.

Creative Law Challengers

Especially, spontaneous (Ps*), introvert-spontaneous (IPs*), intuitive-spontaneous (NPs**) and thinking-spontaneous (TPs**) are creative law challengers. At the type level, even though not statistically significantly, ISTPs, INTPs, ENFPs and ENTPs are clearly overrepresented among creative law challengers.

Correspondingly, as could be concluded according to the meticulous-planners, judgers (Js**), extrovert-judgers (EJs**), sensing-judgers (SJs*), feeling-judgers (FJs*) and extrovert-sensors (ESs*) are a minority among creative law challengers; thinking-judgers clearly, too, even though not significantly. At type level, types preferring judging do not tend to be creative law challengers (Table 8.5).

Conclusions and Lessons to be Learnt

Several relationships between type and creativity orientations were found. In Figure 8.2, a summary of the results is presented. Without repeating the results, it may be concluded that organized sensing personalities prefer mostly meticulous planning orientation. This could be assumed based on the description of the types. They prefer structured, organized, and planned environments.

Table 8.4 Creative Idea Man Compared to the Total Sample

Relationship Between Type and Creativity

Creative Idea Man
compared with
Whole Base
N = 48

ISTJ N = 0 % = 0.00 I = 0.00	**ISFJ** N = 0 % = 0.00 I = 0.00	**INFJ** N = 0 % = 0.00 I = 0.00	**INTJ** N = 1 % = 2.08 I = 0.98 C
ISTP N = 3 % = 6.25 I = 1.18 CCC	**ISFP** N = 2 % = 4.17 I = 1.97 CC	**INFP** N = 0 % = 0.00 I = 0.00	**INTP** N = 2 % = 4.17 I = 1.58 CC
ESTP N = 1 % = 2.08 I = 0.33 C	**ESFP** N = 0 % = 0.00 I = 0.00	**ENFP** N = 13 % = 27.08 I = 2.44** CCCCCCCCCC CCC	**ENTP** N = 6 % = 12.50 I = 1.48 CCCCCC
ESTJ N = 6 % = 12.50 I = 0.64 CCCCCC	**ESFJ** N = 2 % = 4.17 I = 0.39 CC	**ENFJ** N = 2 % = 4.17 I = 1.58 CC	**ENTJ** N = 10 % = 20.83 I = 1.71 CCCCCCCCCC

Print date:0

	N	%	I
E	40	83.33	1.09
I	8	16.67	0.70
S	14	29.17	0.49***
N	34	70.83	1.76***
T	29	60.42	0.94
F	19	39.58	1.10
J	21	43.75	0.75
P	27	56.25	1.35
IJ	1	2.08	0.16*
IP	7	14.58	1.38
EP	20	41.67	1.33
EJ	20	41.67	0.93
ST	10	20.83	0.54*
SF	4	8.33	0.39
NF	15	31.25	2.11**
NT	19	39.58	1.56
SJ	8	16.67	0.41**
SP	6	12.50	0.66
NP	21	43.75	1.92**
NJ	13	27.08	1.55
TJ	17	35.42	0.86
TP	12	25.00	1.10
FP	15	31.25	1.64
FJ	4	8.33	0.49
IN	3	6.25	1.07
EN	31	64.58	1.88***
IS	5	10.42	0.58
ES	9	18.75	0.45**
Sdom	1	2.08	0.09**
Ndom	20	41.67	1.87**
Tdom	21	43.75	1.10
Fdom	6	12.50	0.79

* < .05, ** < .01, *** < .001

Source: Empirical Studies based on The Selection Ratio Type Table, see Granade and Myers, 1987

Note: C = 1 person.
Base total N = 189. Groups are independent.

Table 8.5 Creative Law Challengers Compared to the Total Sample

Relationship Between Type and Creativity

Creative Law Challenger
compared with
Whole Base
N = 53

ISTJ	**ISFJ**	**INFJ**	**INTJ**
N = 4 % = 7.55 I = 1.02 CCCC	N = 1 % = 1.89 I = 0.59 C	N = 0 % = 0.00 I = 0.00	N = 0 % = 0.00 I = 0.00
ISTP N = 6 % = 11.32 I = 2.14 CCCCCC	**ISFP** N = 1 % = 1.89 I = 0.89 C	**INFP** N = 0 % = 0.00 I = 0.00	**INTP** N = 4 % = 7.55 I = 2.85 CCCC
ESTP N = 3 % = 5.66 I = 0.89 CCC	**ESFP** N = 3 % = 5.66 I = 1.07 CCC	**ENFP** N = 9 % = 16.98 I = 1.53 CCCCCCCCC	**ENTP** N = 9 % = 16.98 I = 2.01 CCCCCCCCC
ESTJ N = 8 % = 11.32 I = 0.58 CCCCCC	**ESFJ** N = 1 % = 1.89 I = 0.18 C	**ENFJ** N = 1 % = 1.89 I = 0.71 C	**ENTJ** N = 5 % = 9.43 I = 0.78 CCCCC

	N	%	I
E	37	69.81	0.92
I	16	30.19	1.27
S	25	47.17	0.79
N	28	52.83	1.31
T	37	69.81	1.09
F	16	30.19	0.84
J	18	33.96	0.58**
P	35	66.04	1.58**
IJ	5	9.43	0.71
IP	11	20.75	1.96*
EP	24	45.28	1.45
EJ	13	24.53	0.55**
ST	19	35.85	0.93
SF	6	11.32	0.53
NF	10	18.87	1.27
NT	18	33.96	1.34
SJ	12	22.64	0.56*
SP	13	24.53	1.29
NP	22	41.51	1.82**
NJ	6	11.32	0.65
TJ	15	28.30	0.69
TP	22	41.51	1.82**
FP	13	24.53	1.29
FJ	3	5.66	0.33*
IN	4	7.55	1.30
EN	24	45.28	1.32
IS	13	22.64	1.26
ES	13	24.53	0.59*
Sdom	11	20.75	0.93
Ndom	18	33.96	1.53
Tdom	21	39.62	1.00
Fdom	3	5.66	0.36

Print date:

* < .05, ** < .01, *** < .001

Source: Empirical Studies based on The Selection Ratio Type Table, see Granade and Myers, 1987

Note: C = 1 person.
Base total N = 189. Groups are independent.

Figure 8.2 A Summary of the Relationships between Type and Creativity Orientations

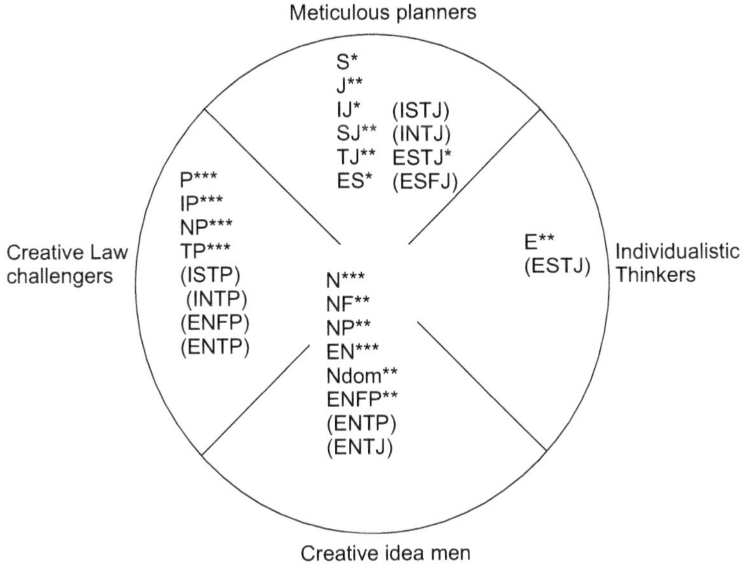

Only extroverts seem to favor individualistic thinking, and consequently, introverts and introvert-sensing types have more temperate behavior. This does not necessarily mean that they have similar views on a matter but they do not express their views so daringly. It must be noted, however, that people may have several roles but they favor some of them.[28] For example, extroverts can be more creative idea men or law challengers but also tending to be individualist thinkers. That is, few people have only individualistic thinker's role in an organization.

Especially, intuition N and intuitive-feeler, intuitive-spontaneous, and especially extrovert-intuitive seem to be associated with creative-idea-man-orientation.

Concerning the 16 types, especially ENFPs like creation, but also ENTPs and ENTJs. In general, the types having N dominance in their preference order tend to be creative idea men.

Entrepreneurial Ps (IPs, TPs, NPs) are strongly oriented to challenge laws and norms. Correspondingly, also the most entrepreneurial personality types tend to challenge the rules. Those preferences and types prefer flexible, spontaneous and changing environments. Studies on entrepreneurial personalities support results found here; entrepreneurs are mostly smaller risk-taking, spontaneous and sensing, or intuitive-spontaneous searching for greater challenges, for example, growth and internalization.[29]

In spite of the limited number of personality types in the sample, there was found clear relationships between the creativity orientations and personality.

Personality preferences and types seem to have clear division of labor in terms of creation and daring. Sensing-judging personalities value traditions and practices found to work. These personalities do not question the prevailing norms that much. Instead, intuitives are pumping up new ideas, and together with spontaneous personalities question the prevailing ideas, and present new ideas.

It is evident that creation and innovation require different personalities, and at different phases of the innovation process, different combinations of personalities should be used in team-building. In the first phase, heterogeneous teams are needed; spontaneous and intuitive have the main role in creations teams, but later, when implementation is concerned, more sensing and judging are in order.

Balanced organizations, teams, and groups — all need creativity orientations. Creative idea men and challengers take care that there will be new ideas as seeds for innovations, individualistic thinkers question non-useful ideas, and meticulous planners put the ideas into work. That is why it is extremely important that organizations should perceive the meaning of difference in building of teams, groups, and units.

The results presented here support the earlier study of Asikainen and Routamaa[30] and McCaulley.[31] However, in this study, a new four-dimension model of creativity orientation was developed in order to find out the distribution of work of types in creativity, in more depth than is done this far.

Notes

1. M. Honkonen and V. Routamaa, 'Psychological Types and Team Work Effectiveness: A Study of Production Teams in Industry', *Proceedings of the Leadership and the Myers–Briggs Type Indicator*, Second International Conference, Washington, 1997.
2. K. Heinäsuo and V. Routamaa, 'Personality Types and Team Roles Hunting Out Effective and Heterogeneous Teams', Proceedings of 8th International Workshop on Teamworking (IWOT 8), Trier, 2004.
3. V. Routamaa and T. Hautala, *Katse Naamion Taa: Itsetuntemuksesta Voimaa*, Leadec-Kustannus, Vaasa, 2009.
4. I. Myers, *Introduction to Type*, Consulting Psychologists Press, Inc., 1992.
5. I. B. Myers and M. H. McCaulley, *A Guide to the Development and Use of the Myers–Briggs Type Indicator*, Consulting Psychologists Press, Inc., 1990.
6. Myers, *Introduction to Type.*
7. Myers and McCaulley, *A Guide to the Development and Use of the Myers–Briggs Type Indicator*.
8. Routamaa and Hautala, *Katse Naamion Taa: Itsetuntemuksesta Voimaa*; H Myers, *Introduction to Type*.
9. V. Routamaa and E. Varamäki, 'Entrepreneurs' Personality and Networking Attitudes', Proceedings of the 25th International Small Business Congress, Sao Paulo, 1998; D. Keirsey and M. Bates, *Please Understand Me*, Prometheus Nemes Books, Del Mar, 1984.
10. Keirsey and Bates, *Please Understand Me*.
11. J. C. Carland, J. W. Carland and R. C. Higgs, 'Student Performance in Management: Is It a Function of Learning?' Proceedings of the Institute of Management Sciences Southeastern Chapter, 1993, 289–91.
12. Keirsey and Bates, *Please Understand Me*.
13. V. Routamaa, 'An Entrepreneur – A Singular Character and a Challenge for Educators', in *Work in Process* (eds P. Malinen and K. Paasio), Turku School of Economics, Series C 1, Esa Print, Tampere, 2008.
14. Myers and McCaulley, *A Guide to the Development and Use of the Myers–Briggs Type Indicator*'.

15. N. D. Gryskiewicz and W. L. Tullar, 'The Relationship between Personality Type and Creativity Style among Managers', *Journal of Psychological Type*, 32, 1995, 30–35.
16. J. Van Rooyen, 'Creativity: An Important Managerial Requirement. A South African Perspective', Proceedings of the Myers–Briggs Type Indicator and Leadership: An International Research Conference, 1994.
17. G. C. Carne and M. J. Kirton, 'Styles of Creativity: Test-score Correlations between Kirton Adaption-Innovation Inventory and Myers–Briggs Type Indicator', *Psychological Reports*, 50, 1982, 31–36.
18. Ibid.
19. M. H. McCaulley, 'The Myers–Briggs Type Indicator and Leadership', in *Measures of Leadership: Center for Creative Leadership* (eds Clark and Clark), Greensboro, 1990.
20. V. Asikainen and V. Routamaa, 'The Relationship between the MBTI and the Creativity Orientations of Managers', Proceedings of the Leadership and the Myers–Briggs Type Indicator, Second International Conference, Washington, 1997.
21. Ibid.
22. McCaulley, 'The Myers–Briggs Type Indicator and Leadership'.
23. Asikainen and Routamaa, 'The Relationship between the MBTI and the Creativity Orientations of Managers'.
24. V. Asikainen, 'Personality Types and Creativity Orientations of Managers', Unpublished Licentiate Thesis, University of Vaasa, 1996.
25. Asikainen and Routamaa, 'The Relationship between the MBTI and the Creativity Orientations of Managers'.
26. R. E. Byrd, *C & RT*, San Diego: University Associates, 1986.
27. J. G. Granade and I. B. Myers, 'The Selection Ratio Type Table PC Program', Center for Applications of Psychological Type, Gainesville, 1987.
28. Heinäsuo and Routamaa, 'Personality Types and Team Roles Hunting Out Effective and Heterogeneous Teams'.
29. V. Routamaa and A. Miettinen, 'Awareness of Entrepreneurial Personalities: A Prerequisite for Entrepreneurial Education', NCGE Working Paper Series, National Council for Graduate Entrepreneurship, Birmingham, 2007; V. Routamaa and A-L. Rissanen, 'Entrepreneurial Identity and Personality Types', Proceedings of the 8th International Conference on Global Business

& Economic Development, Guadalajara, 2004; V. Routamaa, J. Vesalainen and E. Pihlajaniemi, 'Meeting the Challenges of Export Markets: Entrepreneurs' Personality and International Orientation', Proceedings of the 7th ENDEC World Conference, 1996.

30. Asikainen and Routamaa, 'The Relationship between the MBTI and the Creativity Orientations of Managers'.
31. McCaulley, 'The Myers–Briggs Type Indicator and Leadership'.

9

Diversity and Innovative Performance in Teams: The Role of Conflict-Management Styles, Team and Leadership Identity

Aline D. Masuda and Konstantinos C. Kostopoulos

Dr. Randy Pausch, a professor in computer engineering from Carnegie Mellon University, is well-known for the last lecture he gave before he died of pancreatic cancer in 2008. In this lecture, he explained why he turned down a chance to fulfill his childhood dream of working at Walt Disney Imagineering to continue working as a professor at Carnegie Mellon University where he could help his students reach their own childhood dreams.[1]

Dr. Pausch also explained that it was after turning down the offer to work at Imagineering that he decided to create a course called 'Building virtual worlds' where he brought together a team of students majoring in art design, drama, and computer science to work on a two-week project developing entertainment software programs. The course was so successful that students, professors from other courses, and even parents lined up outside his classroom to watch students present their creations.

Dr. Pausch knew how to take advantage of diversity to promote creativity in teams, and for this reason, his classes were so successful. Research has already shown that teams comprised of people with diverse backgrounds are more likely to produce a creative output than homogeneous teams.[2] *However, studies have also shown that diversity may not always lead to creativity, and that several factors can moderate this relationship.*[3] *Specifically, while heterogeneous teams are more likely to take advantage of their diverse pool of resources and produce more creative and innovative work,*[4] *these teams are also more likely to face emotional conflict, and experience different forms of misunderstandings.*[5] *In this paper, we review the literature on the relationship between team-diversity, team-creativity, and innovation. Furthermore, we propose a model that explains how team-conflict management styles, and team-and leadership identity could moderate the relationship between team-diversity, team-creativity, and innovation. In particular, we argue that these factors will moderate the relationship by influencing the level of conflict in teams.*

The paper is structured as follows. First, we report the latest research that examines the relationship between team-diversity with creativity and innovation. Second, we review the literature on conflict and team-diversity, and explicate that conflict is a particular challenge experienced by heterogeneous groups. Third, we report how conflict-management styles, team-and leadership-identity could potentially strengthen the effect of team-diversity on creativity and innovation. We primarily use social identity theory[6] and information process theory[7] to develop our arguments.

Team-Diversity and Creative Performance

Several empirical studies have shown the benefits of team-diversity on performance, creativity, and innovation.[8] While team-creativity can be defined as 'the production of novel and useful ideas concerning products, services, processes, and procedures by a team of employees working together',[9] team-innovation has been defined as 'the intentional introduction and application within a team, of ideas, processes, products or procedures new to the team, designed to significantly benefit the individual, the team, the organization, or wider society'.[10]

Team-creativity can be observed in Dr. Pausch's classroom described above. In his last lecture, Dr. Pausch describes an original video game called 'Hello World', which was created by a heterogeneous team in his class. In this video game, users were able to use their own imagination to create and destroy their own 'world'. This was an example of team-creativity as the output produced by the team was novel, and had a specific purpose (e.g., to entertain). If the team introduced an idea that already existed in one context to create something useful (for example, besides creating the video game, the team of students could adapt this game so that it would not only be entertainment but also educational), then the team-output would be innovative in addition to being creative.

In this chapter, we review the literature on both team-innovation and creativity. However, when stating our propositions, we will focus on innovative performance as it includes creativity. We could argue that one of the antecedents of innovative performance in Dr. Pausch's classroom, was the type of students or the type of diversity that comprised his work teams. Researchers have categorized diversity into two broad types: diversity that is easily visible such as gender, race, age, and ethnicity, and diversity that are not readily visible, such as culture and attitudes, or cognitive styles. Some researchers define visible diversity as surface-level diversity (e.g., demographic) and non-visible diversity as deep-level diversity (e.g., attitudes or cultural values).[11] Others similarly categorize diversity as 'observable individual differences' and 'underlying attributes'.[12] In the example above, Dr. Randy Pausch created teams comprised of members with underlying attributes that may not be visible such as cognitive styles, knowledge, personality, and values which can be unique features of individuals, who are majoring in art design, drama and computer sciences, i.e., deep-level diversity. However, these teams were probably comprised of individuals having different age, gender and perhaps ethnicity, i.e., surface-level diversity.

Research has shown that deep-and surface-level diversity in teams moderate the diversity-innovation relationship.[13] Specifically, in a quantitative review of 35 empirical studies examining diversity and team-performance outcomes, Horwitz and Horwitz[14] showed that deep-level diversity, defined as task diversity, leads to higher team-innovation. However, surface-level diversity—such as age, gender and race—did not relate to team-innovation. Horwitz and Horwitz[15] explained these differences by drawing from information process theory,[16] social identity theory,[17] and the similarity-attraction paradigm.[18] The information-processing hypothesis explains why task-related diversity leads to innovation/creativity. Specifically, cognitive diversity or expertise diversity leads to innovation and creative performance, because individuals contribute to the group with their unique knowledge that is essential to develop a creative output. For example, a team comprised of artists and computer engineers are more capable of creating an innovative entertainment software program than a team comprised only of computer engineers because the artists could provide the expertise related with storytelling while the computer engineers could provide the technical know-how.

While the information-processing theory explains how deep-level diversity can lead to positive outcomes in terms of innovation/creativity, the similarity (attraction and social identity) theory explains why surface-level heterogeneity in a group, can be an impediment in team-creativity/innovation. According to social identity theory and self-categorization theory,[19] individuals who belong to a certain category or group (e.g., age or gender) tend to have favorable perceptions toward people who belong to their in- group compared with people who do not belong to the in-group.[20] For example, scientists tend to have favorable perceptions towards scientists more so than towards business professionals. This is because categorization can lead to stronger identification with the group one belongs to, e.g., gender, race, and professional affiliation. That is, people develop a social identity, which is defined as people's sense of self, derived from their affiliation in groups that transmit and instill knowledge, beliefs, values, attitudes, and traditions.[21] Because individuals develop a strong emotional attachment towards their social groups, they are likely to react emotionally towards anyone who may not belong to their in-group and who may threaten their social identity. For this reason, heterogeneous teams are likely to experience more emotional conflict. As Horwitz and Horwitz[22] argued, having biodiversity in teams leads to the categorization and creation of subgroups comprised of similar individuals, which consequently leads to in-group and out-group bias, making the group environment more propitious to interpersonal conflict between team members, who belong to different categories.

The similarity-attraction paradigm also explains, why interpersonal conflict between heterogeneous teams comprised of surface-level diversity is likely to occur. This paradigm explains that when people are given the option to choose whom to work with, they tend to choose to work with people who are similar to them. In addition, people are also more likely to be satisfied when they are working with individuals with a similar background.[23] Thus, based on Horwitz and Horwitz's[24] study and the aforementioned theories, while deep-level diversity can lead to creativity in teams 'the beneficial linkage between bio-demographic diversity and team-performance suggested in the team-literature has been overstated particularly considering the extremely weak to nonexistent relationship demonstrated in this meta-analysis'.

Despite these findings, surface-and deep-level diversity cannot be easily separated. For example, cultural diversity is a unique type of diversity that encompasses both surface-level factors and deep-level diversity.[25] Values and beliefs are deep-level aspects. Other aspects of culture such as language, and symbols make cultural diversity more observable.

Because cultural diversity has both surface-and deep-level factors, researchers have argued, based on the information-processing and social identity theories as well as similarity attraction hypothesis, that cultural diversity could have both positive and negative outcomes to team-performance.[26] Recently, Stahl et al.[27] published a quantitative review including 108 empirical studies examining the effects of cultural diversity in team-creativity. They proposed a conceptual framework that explains how cultural diversity leads to team-performance via several mediators. One of these mediators was team-creativity. Their review showed an effect size of 0.16 between diversity and creativity. Contrary to Horwitz and Horwitz's study,[28] there were no significant differences in the effect size when looking at the relationship between surface-and deep-level diversity with creativity. Based on the points made above, we argue that both deep-and surface-level diversity are potential antecedents of innovative performance in work teams.

Proposition 1: Deep-level and surface-level diversities are positively related with innovative performance in work teams.

Task Conflict as a Mediator in the Relationships between Team-Diversity, Creativity, and Innovation

As previously stated, research has shown that heterogeneous work teams are more likely to experience intra-team conflict and misunderstandings.[29] Intra-team conflict occurs when 'team members hold discrepant views or have interpersonal incompatibility'.[30] Researchers have shown that relationship-based conflicts, which are conflict-based on differences in personality, is detrimental to team-performance and morale[31], while task-based conflict, which occur when team-members disagree with allocation of resources, procedures, or differences in task interpretations,[32] can have beneficial effects to the group if handled appropriately.[33]

Despite the contentions that conflict is related to innovation in teams, a recent meta analytical review of 104 empirical studies examining the antecedents of team-innovation showed a slightly positive relationship between task conflict and team-innovation, and a negative relationship between relationship conflict and team-innovation.[34] However, the relationships were not significant. Hulsheger et al.,[35] explained that there are possible moderators that might better explain under which conditions conflict predicts innovation and creativity in teams. This is congruent with De Dreu's[36] contention that conflict may be beneficial only under specific circumstances. For example, a study conducted by Farh, Lee, and Farh[37] showed that, in moderate amounts, task conflict can lead to team-creativity. Specifically, they conducted a study using 71 information technology teams, and found a curvilinear relationship between task conflict and team-creativity, with team-creativity being highest at moderate levels of task conflict and lowest at highest or lowest level of task conflict. Interestingly, they also found that the time when team-conflict occurred moderated this relationship. That is, task conflict that occurred in the later stages of team-development did not have an effect on creative performance. Instead, the curvilinear effect was found only when task conflict happened in the earlier stages of team-development.

Another condition under which task conflict may promote innovation is the extent to which team members feel comfortable in expressing their doubts.[38] Further, research has shown that the way team members approach their conflict influences whether the conflict is constructive or destructive.[39] The ability to manage conflict constructively may give heterogeneous teams some leverage to take advantage of their diverse resources to create and innovate. Below, we summarize the research on team-conflict management styles, and argue that managing team-conflict could be a condition under which conflict leads to creativity. Specifically, we contend that the way teams manage conflict is an important condition that allows team members to exploit the important resources that originate from a diverse team, in order to produce creative or innovative outcomes.

Conflict-Management Styles as a Moderator in the Relationships between Team-Diversity, Creativity, and Innovation

As previously stated, in the context of teamwork, conflicts often occur due to differences in members' viewpoints regarding the content of the group task (i.e., task conflicts) or due to interpersonal incompatibilities and dislikes (i.e., relationship conflicts).[40] The way through which teams *manage* conflict determines conflict emergence and evolution, and whether it will eventually impede or inversely enhance the productivity and creative performance of organizational teams.

In their seminal framework of team processes, Marks and her colleagues[41] categorize conflict management as one of the key interpersonal processes that a team should foster. They argue that conflicts, either task or relationship, can be managed in a pre-emptive (i.e., before they actually occur) or in a reactive manner (i.e., after they have emerged). In both cases, the development of group norms and rules for *cooperative* rather than *competitive* approaches to conflict management is of utmost importance, in order to offset the detrimental effects or realize the potential benefits of conflicts.[42]

In line with Deutsch's[43] theory of cooperation and competition, team members may adopt either a cooperative style in managing conflicts through emphasizing their common goals, or a competitive style by focusing on achieving personal goals in conflict situations. Cooperative style entails a high level of concern for others besides the self, and has been characterized as a win-win, collaborative and problem-solving approach that seeks to resolve conflicts for mutual benefit. By contrast, competitive style involves a high level of concern for personal interests at the expense of the other parties, and has been described as a win-lose and controlling approach toward managing conflict. Scholars have used the cooperative-competitive conflict management dichotomy in organizational contexts,[44] demonstrating the effects that cooperative or competitive styles may have on several group dynamics such as trust and decision making quality, as well as on team outcomes such as members' satisfaction and group efficiency.

Building further on the effects of conflict management styles, we hereby argue that cooperative and competitive styles can play an important moderating role in the relationship between (surface- and deep-level) diversity and creativity or innovation in work teams. Specifically, it has been contemplated that a cooperative approach to conflict promotes social integration within teams,[45] creating a climate of trust, cohesion, and freedom to express different views and opinions.[46] This cooperative climate is of particular usefulness in heterogeneous teams, where differences and disagreements regarding group tasks and procedures are most likely to emerge. In such groups, therefore, cooperative conflict-management should provide a platform for constructive expression of opposing views, identification of mistakes, and collaboration to handle differences and accomplish common goals. Feeling more socially integrated with their teams, members can thus consider each other's demographic, cultural, or cognitive discrepancies open-mindedly, understand them, and capitalize more on the creativity and knowledge creation opportunities that they offer.[47] Hence, in situations where conflicts are handled cooperatively, heterogeneous teams are likely to experience superior creativity and innovation by motivating team members to work together to solve complex problems and, in turn, benefiting from their diverse ideas and novel insights.

Team members, on the other hand, may also manage conflicts in a competitive manner, emphasizing their personal goals and creating 'I-win-you-lose' situations. Such a competitive approach tends to inhibit intra-group communication and generate feelings of distrust among team members.[48] These detrimental effects can be even more serious in teams of increased diversity. In such settings, a competitive conflict-management style may further amplify the existing demographic or cultural differences among individual members, reinforcing possible 'us-vs-them' attitudes between different subgroups, while simultaneously 'neutralizing' potential information benefits that could emerge because of diversity. By focusing on achieving personal goals and competing with each other, team members cannot share a sense of belonging and identity with their team,[49] thus failing to create a psychologically safe environment that allows open-minded discussions of different views and opinions. Consequently, if heterogeneous teams adopt a competitive approach towards managing conflict, then their creative performance will likely decrease, due to distractions in the communication and exchange of ideas, solutions, and resources. This argument is consistent with recent findings such as of Desyvilya, Somech, Lidgoster,[50] who conducted a study of 77 work teams from a high technology company, and showed that team identity was positively related with innovation, and that a team's integrating conflict management style, which entails collaboration, mediated this relationship. Interestingly, their study also showed that task conflict was positively related to this type of conflict-management style.

Hence, we suggest that under a cooperative conflict-management style, team-diversity will be positively related with team-creativity, while, in contrast, under a competitive conflict-management style, this relationship becomes negative.

Proposition 2: The conflict-management style will moderate the relationship between team-diversity and innovative performance. Under cooperative conflict-management style, the relationship between diversity and innovative performance will be positive. Under competitive conflict-management style, the relationship between diversity and innovative performance will be negative. This moderation will be mediated by types of conflict.

Team-Identity as a Moderator in the Relationships between Team-Diversity, Creativity, and Innovation

As previously stated, group categorization is more likely to occur in heterogeneous work teams composed of members with different educational backgrounds (i.e., deep-level diversity) and different gender or age (i.e., surface-level diversity) as compared with homogeneous teams.[51] As such, under these conditions, interpersonal conflict is likely to occur under threats to social identity. For example, group categorization could take place within a multicultural team comprised of members from Latin American and European countries, observing the development of two sub-groups with members from each geographic location. The same can be suggested for a group comprised of people from distinct professional backgrounds — e.g., engineers and artists as in the example above. For this reason, as we mentioned above, although heterogeneous work teams are more likely to provide resources necessary to create innovative work, they are also more likely to experience process loss by encountering more conflict due to the potential development of subgroups.[52]

Despite the fact that heterogeneous teams are, by definition, comprised of individuals representing different social groups thus reflecting different social identities, such teams could also develop a unique sense of team-identity. According to Mael and Ashforth,[53] team-identification refers to how team members consider team goals as their own, and may feel 'psychologically intertwined with the group's fate'. Earley and Mosakowsky[54] defined this shared meaning system as a 'common sense of entitativity', while Van Der Vegt and Bunderson,[55] described it as an 'emotional significance that members of a given group attached to their membership in that group'.

Team-identification has been proposed to have important implications for team-performance.[56] For example, Lembke and Wilson,[57] explain that if team members in heterogeneous teams develop a strong sense of team-identity, such that their membership in the team could be more self-defining compared with membership in other social groups—e.g., nationality, profession, gender—, then they would be better prepared to manage conflict, because this sense of team-identification unifies and motivates team members to work towards a common purpose. Further, if team members have a strong sense of team-identity, the team's success or failure becomes their own personal interest.[58]

There has been some empirical evidence showing that team-identification can lead to better conflict management. For instance, Somech, Desyvilia and Lidogostar,[59] reported that task interdependence were related to cooperative style of managing conflict, when teams had a strong team-identity. The cooperative style of managing team-conflict, in turn, related to team-performance. Further, research conducted with geographically distributed teams showed that teams with a strong sense of team- identification were less likely to experience task conflict.[60] This finding was only observed in geographically distributed teams. Finally, Han and Harms[61] found that team-identification led to lower level of relationship and task conflict in research and development (R&D) teams, and that this relationship was mediated by trust.

Based on the above discussion, we argue that team identity can provide the team with the emotional significance that they need, in order to promote an environment that help members to deal with disagreements. By shifting the focus from interpersonal differences to commonality (e.g., common goal), group members will be less likely to engage in interpersonal conflict, and will be better prepared to handle disagreements related with the task. This, in turn, would enable group members to take advantage of the diversity of resource, such as unique knowledge and points of views, in order to produce innovation or creativity. This is in line with previous studies, that showed a relationship between team-identity with innovation and creativity via conflict management.[62] Specifically, Paulsen et al.[63] examined 34 work teams from an R&D organization, and found that team-identity was positively related to cooperation in teams, and cooperation in teams was positively related to innovation. Desyvilya, et al.,[64] proposed a model where they define team-identity as the team's capacity to mitigate the adverse impact of relationship conflict, and its capability to maximize the potential gains of task conflict.

Taken together, we propose that team-identity will strengthen the relationship between team-diversity and creativity by reducing relationship conflict and maintaining an optimal level of task conflict.

Proposition 3: Team-identity moderates the relationship between team diversity and innovative performance. Under high team-identity, the relationship between team-diversity and innovative performance will be positive. Under low team-identity, the relationship between team-diversity and innovative performance will be nil. This moderation will be mediated by task conflict and relationship conflict.

Leadership Identity as a Moderator in the Relationships between Team Diversity, Creativity, and Innovation

It is well acknowledged by researchers that team leadership plays a fundamental role in team performance, viability, and well-being.[65] Researchers also support that leadership effectiveness is a function of both situation and leadership attributes.[66] According to the functional leadership model,[67] effective leaders are able to identify and fulfill critical team functions under key situations in order to ensure team success.[68] Being in a heterogeneous team is a unique situation that requires unique team-leadership attributes. As such, based on this model, team leaders should play a significant role in promoting innovation when working in a heterogeneous team, provided that they fulfill a specific function that would foster team innovative-performance.

As previously stated, conflict management and the development of a strong team-identity are critical functions that could be fulfilled by leaders in heterogeneous teams in order to promote team-innovative performance. We contend that team leaders who are capable of managing conflict in heterogeneous groups may be more adept at avoiding team process losses by promoting the conditions that facilitate the exchange of ideas among team members.

A unique leadership-attribute that could influence the leader's ability to help team members manage conflict and develop a strong team identification is leadership identity. Based on social identity theory of leadership,[69] leadership identity can be instrumental in helping team members perform such functions, because it influences leadership emergence and perceptions of leadership effectiveness. In particular, social identity analysis of leadership suggest that team members with characteristics prototypical of their team, will be more likely to be trusted and perceived as leaders in such a team.[70] In newly formed teams, the prototypicality does not necessarily mean that the leader exerts influence over team members. However, with time, prototypical leaders become influential through depersonalized social-attraction processes. Specifically, team members tend to like prototypical team members more than non-prototypical members, and they also tend to perceive prototypical members as more charismatic. By being more socially attractive, prototypical members are more able to convince team members to comply with their ideas. Such members are also perceived to be more effective. This is important, as research has shown that charismatic leaders promote innovation by fostering a sense of team identity, and encouraging participation and exchange of ideas.[71]

Based on this line of reasoning, in heterogeneous groups, a prototypical leader should be a leader that identifies with different social groups represented by members of the group. For example, based on this argument, team members in a team comprised of artists and engineers will likely perceive someone as an effective leader, if this person has both artistic and engineering capabilities or can, at least, identify with both groups. Research on biculturalism has shown that people can develop multiple identities, and those who are able to identify with multiple cultures are more likely to adapt in a cross-cultural environment. Specifically, biculturalism is related with heightened level of cognitive/integrative complexity[72] and there is higher likelihood that individuals will adopt and endorse certain values.[73] Cultural identities are defined as 'a sense of solidarity with the ideals of a given cultural group and to the attitudes, beliefs, and behaviors manifested toward one's own (and other) cultural groups'.[74]

If bicultural individuals are more likely to adopt higher levels of integrative complexity, then individuals who identify with multiple disciplines or other social identities, could also be more likely to develop some sensibility and abilities that are essential, when leading multidisciplinary teams. By identifying with multiple disciplinary groups, a leader may be viewed as influential or charismatic in a multidisciplinary team. Through identifying with multiple generations, a leader may be viewed as more influential and charismatic in a team comprised of individuals from different generations. The same can be supported with other types of diversity, e.g., gender. By being more influential in teams, such leaders may be more capable to create an environment, where teams adopt a cooperative style to manage conflict. Further, such leaders could also be more likely to help the team to develop a strong overarching team identity, that may guide member's activities toward a common (team) pattern of behavior. Based on the above discussion, we contend that leadership identity will indirectly influence innovative performance by influencing team identity and the team conflict management style.

Proposition 4: Leadership identity will indirectly influence innovative performance in heterogeneous teams via improvements in team identity and team's conflict-management style.

Figure 9.1 illustrates our propositions.

Figure 9.1 A Conceptual Model of Team Innovative Performance

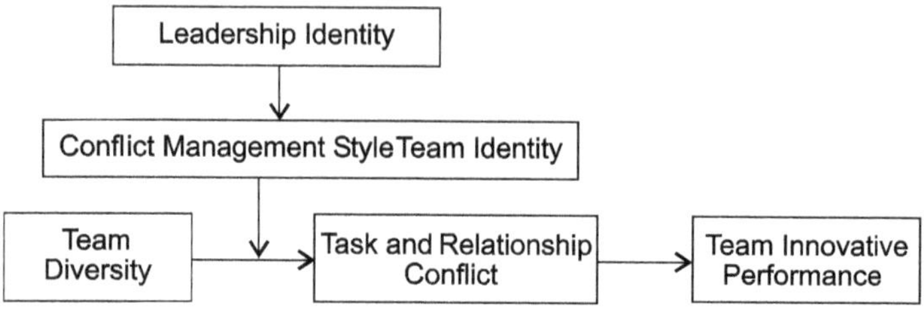

Conclusion

In this chapter, we reviewed and synthesized the empirical evidence linking team diversity—both deep-level and surface-level diversity — with team innovative-performance. We also provided a conceptual model proposing moderators and mediators in the diversity-innovation relationship. A key mediating condition in this relationship is conflict. We argue that moderate levels of task conflict can be beneficial to team innovation-performance, and that task conflict could lead to innovation if managed appropriately. Furthermore, we identified conflict-management styles and team identity as moderators in the team diversity and team innovative-performance relationship. Specifically, teams with a cooperative conflict-management style and a strong sense of identity would be more likely to manage both task and relationship conflict, so that they can derive important innovative-enhancing benefits. We develop our conceptual model based on social identity theory[75] and information process theory.[76]

We also suggested that team leadership is an important antecedent of team innovative-performance. Specifically, we argued that leaders who are able to identify with different social groups, represented in heterogeneous teams, would be capable to exert increased influence in these team settings. For example, such leaders will be more capable to help the team develop a strong team identity, and to collaborate when confronted with interpersonal conflicts. This will allow the team to take advantage of the diversity of knowledge and expertise present in heterogeneous teams. As shown in the example given at the beginning of the chapter, the creativity that was developed in Dr. Randy Pausch's class would not have occurred without Dr. Pausch's initiative to create teams that are purposely multidisciplinary. For this, Dr. Pausch's identification and focus on diverse disciplines and backgrounds was useful. We believe that leaders who identify with multiple disciplines or cultures can better foster a strong team-identity, and help the team members to develop an effective conflict-management style, which are essential conditions for innovative performance. This could be done in the beginning of a team's life cycle, where leaders choose members from different disciplines and functional divisions, but also in the middle of a team's cycle, as heterogeneous self-management teams could rely on an informal leader with prototypical characteristics to help them in managing conflict. By overcoming relationship conflict and harnessing appropriate levels of task conflict, we believe that heterogeneous teams can better work towards innovation.

Notes

1. Lecture by Randy Pausch, http://www.cs.cmu.edu/~pausch/ (accessed 29 August 2013).
2. G. K. Stahl, M. L. Maznevzki, A. Voigt and K. Jonsen 'Unraveling the Effects of Cultural Diversity in Teams: A Meta-Analysis of Research on Multicultural Work Groups', *Journal of International Business Studies*, 41, 2010, 690–709.
3. S. K.Horwitz and I. B. Horwitz, 'The Effects of Team Diversity on Team Outcomes: A Meta-Analytic Review of Team Demography', *Journal of Management*, 33, 2007, 987–1015.
4. Stahl, Maznevzki, Voigt, and Jonsen, 'Unraveling the Effects of Cultural Diversity in Teams: A Meta-Analysis of Research on Multicultural Work Groups'.
5. M. A. Van Glinow, D. L. Shapiro, and J. M. Brett, 'Can We Talk, and Should We? Managing Emotional Conflict in Multicultural Teams', *Academy of Management Review*, 29(4), 2004, 578–92; K. Behfar, M. Kern and J. Brett, 'Managing Challenges in Multicultural Teams' in *Research on Managing Groups and Teams: National Culture and Groups* (eds E.A.M. Mannix, M.

Neale, & Y. Chen), 9, Oxford: Elsevier Science Press, 2006, 239–69.

6. H. Tajfel and J. C. Turner, 'The Social Identity Theory of Intergroup Behavior' in *Psychology of Intergroup Relations* (eds S. Worchel and W. G. Austin), Nelson-Hall, Chicago, 1986, 7–24.
7. D.G. Ancona and D.F. Caldwell, 'Demography and Design: Predictors of New Product Team Performance', Organization Science, 3(3), Cox, 1992, 321–41; P.M. Blau, *Inequality and Heterogeneity: A Primitive Theory of Social Structure*, Free Press, New York, 1977, Cox, *Cultural Diversity in Organizations: Theory, Research, and Practice*; T. H. Cox, *Cultural Diversity in Organizations: Theory, Research, and Practice*, Benett-Koehler Publishers, San Francisco, 1994.
8. J. Galbraith, *Designing the Global Corporation*, Jossey Bass, San Francisco, 2000; Stahl, Maznevzki, Voigt and Jonsen, 'Unraveling the Effects of Cultural Diversity in Teams: A Meta-Analysis of Research on Multicultural Work Groups'; Horwitz and Horwitz, 'The Effects of Team Diversity on Team Outcomes: A Meta-Analytic Review of Team Demography'; P. McLeod and S. Lobel, 'The Effects of Ethnic Diversity on Idea Generating Groups', Academy of Management Best Paper Proceedings, 1992, 227–31; U. Hulsheger, N. Anderson, J. Salgado, 'Team-Level Predictors of Innovation at Work: A Comprehensive Meta-Analysis Spanning Three Decades of Research', *Journal of Applied Psychology*, 94, 2009, 1128–45.
9. S. Shin and J. Zhou, 'When is Educational Specialization Heterogeneity Related to Creativity in Research and Development Teams? Transformational Leadership as a Moderator', *Journal of Applied Psychology*, 92, 2007, 1709–21.
10. M.A. West and M. Wallace, 'Innovation in Health Care Teams', *British Journal of Social Psychology*, 21, 1991, 303–15.
11. D.A. Harrison, K.H. Price and M. P. Bell, 'Beyond Relational Demography: Time and the Effects of Surface and Deep-level Diversity on Work Group Cohesion', *Academy of Management Journal*, 41, 1998, 96–107.
12. F.J. Milliken and L.L. Martins, 'Searching for Common Threads: Understanding the Multiple Effects of Diversity in Organizational Groups', *Academy of Management Review*, 21, 1996, 402–33.
13. Horwitz and Horwitz, 'The Effects of Team Diversity on Team Outcomes: A Meta-Analytic Review of Team Demography'.

14. Ibid.
15. Ibid.
16. T. Cox and S. Blake, 'Managing Cultural Diversity: Implications for Organizational Competitiveness', *Academy of Management Executive*, 5(3), 1991, 45–56; Ancona and Caldwell, 'Demography and Design: Predictors of New Product Team Performance'; Blau, *Inequality and Heterogeneity: A Primitive Theory of Social Structure*; Cox, *Cultural Diversity in Organizations: Theory, Research, and Practice*; S. E. Jackson, 'Team Composition in Organizational Settings: Issues in Managing an Increasingly Diverse Workforce', in *Group Process and Productivity* (eds S. Worchel, W. Wood and J. A. Simpson), Sage, Newbury Park, 1992, 138–73; R. Katz, 'The Effects of Group Longevity on Project Communications and Performance', *Administrative Science Quarterly*, 27(1), 1982, 81–104; W. E. Watson, K. Kumar and L. K. Michaelsen, 'Cultural Diversity's Impact on Interaction Process and Performance: Comparing Homogeneous and Diverse Task Groups', *Academy of Management Journal*, 36(3), 1993, 590–602; K. E. Weick, 'The Social Psychology of Organizing', Addison–Wesley, Massachusetts, 1979.
17. H. Tajfel, 'Social Psychology of Intergroup Relations', Annual Review of Psychology, 3, 1982, 1–39.
18. J. Lincoln and J. Miller, 'Work and Friendship Ties in Organizations: A Comparative Analysis of Relational Networks', *Administrative Science Quarterly*, 24, 1979, 181–99; D. Byrne, G. Clore and P. Worchel, 'The Effect of Economic Similarity — Dissimilarity as Determinants of Attraction', *Journal of Personality and Social Psychology*, 4, 1966, 220–24.
19. J. C. Turner, 'Rediscovering the Social Group: A Self-categorization Theory', Basil Blackwell, Oxford, 1987.
20. Tajfel, 'Social Psychology of Intergroup Relations'.
21. Tajfel and Turner, 'The Social Identity Theory of Intergroup Behavior'.
22. Horwitz and Horwitz, 'The Effects of Team Diversity on Team Outcomes: A Meta-Analytic Review of Team Demography'.
23. Lincoln and Miller, 'Work and Friendship Ties in Organizations: A Comparative Analysis of Relational Networks'; Byrne, Clore and Worchel, 'The Effect of Economic Similarity — Dissimilarity as Determinants of Attraction'.

24. Horwitz and Horwitz, 'The Effects of Team Diversity on Team Outcomes: A Meta-Analytic Review of Team Demography'.
25. Stahl, Maznevzki, Voigt, and Jonsen, 'Unraveling the Effects of Cultural Diversity in Teams: A Meta-Analysis of Research on Multicultural Work Groups'.
26. Ibid.
27. Ibid.
28. Horwitz and Horwitz, 'The Effects of Team Diversity on Team Outcomes: A Meta-Analytic Review of Team Demography'.
29. M. A. V. Glinow, D.L. Shapiro and J. M. Brett, 'Can We Talk, and Should We? Managing Emotional Conflict in Multicultural Teams', *Academy of Management Review*, 29, 2004, 578–92; Behfar, Kern and Brett, 'Managing Challenges in Multicultural Teams'.
30. L. E. Weingart and K.A. Jehn, 'Manage Intra-Team Conflict through Collaboration' in *Handbook of Principles of Organizational Behavior: Indispensible Knowledge for Evidence-based Management* (ed. E.A. Locke), 2, 2009, 327–46.
31. C. K. W. De Dreu and L. R. Weingart, 'Task Versus Relationship Conflict and Team Effectiveness: A Meta-Analysis', *Journal of Applied Psychology*, 88, 2003, 741–49.
32. Ibid.
33. Weingart and Jehn, 'Manage Intra-Team Conflict through Collaboration'.
34. Hulsheger, Anderson and Salgado, 'Team-level Predictors of Innovation at Work: A Comprehensive Meta-Analysis Spanning Three Decades of Research'.
35. Ibid.
36. C. K. W. De Dreu, 'The Virtue and Vice of Workplace Conflict: Food for Pessimistic Thought', *Journal of Organizational Behavior*, 29, 2008, 5–18.
37. J. Farh, C. Lee and C. Farh, 'Task Conflict and Team Creativity: A Question of How Much and When', *Journal of Applied Psychology*, 95, 2010, 1172–80.
38. K. Lovelace, D. L. Shapiro and L. R. Weingart, 'Maximizing Cross-functional New Product Teams' Innovativeness and Constraint Adherence: A Conflict Communications Perspective', Academy of Management Journal, 44, 2001, 779–93.
39. O. B. Ayoko, C. E. J. Hartel and V. J. Cullen, 'Resolving the Puzzle of Productive and Destructive Conflict in Culturally

Heterogeneous Work Groups: A Communication Accommodation Theory Approach', *International Journal of Conflict Management*, 13, 2002, 165–87.

39. C. De Dreu and E. van de Vliert, 'Using Conflict in Organizations', Sage, Beverly Hills, 1997.
40. K. A. Jehn, 'A Multimethod Examination of the Benefits and Detriments of Intragroup Conflict', *Administrative Science Quarterly*, 40, 1995, 256–82.
41. M. A. Marks, J. E. Mathieu and S. J. Zaccaro, 'A Temporally-based Framework and Taxonomy of Team Processes', *Academy of Management Review*, 26, 2001, 358–76.
42. D. Tjosvold, 'Implications of Controversy Research for Management', *Journal of Management*, 11, 1985, 21–37.
43. M. Deutsch, 'The Resolution of Conflict', Yale University Press, New Haven, 1973.
44. D. Tjosvold, 'Bridging East and West to Develop New Products and Trust: Interdependence and Interaction between a Hong Kong Parent and North American Subsidiary', *International Journal of Innovation Management*, 3, 1999, 233–52; D. Tjosvold, C. Hui and Z. Yu, 'Conflict Management and Task Reflexivity for Team In-Role Performance and Extra-Role Performance in China', *International Journal of Conflict Management*, 14, 2005, 141–63; A. Somech, H. S. Desivilya and H. Lidogoster, 'Team Conflict Management and Team Effectiveness: The Effects of Task Interdependence and Team Identification', *Journal of Organizational Behavior*, 30, 2009, 359–78; P. S. Hempel, Z. H. Zhang and D. Tjosvold, 'Conflict Management between and within Teams for Trusting Relationships and Performance in China', *Journal of Organizational Behavior*, 30, 2008, 41-65.
45. Tjosvold, Hui and Yu, 'Conflict Management and Task Reflexivity for Team In-Role Performance and Extra-Role Performance in China'.
46. Somech, Desivilya and Lidogoster, 'Team Conflict Management and Team Effectiveness: The Effects of Task Interdependence and Team Identification'.
47. C. K. W. De Dreu, 'When Too Little or Too Much Hurts: Evidence for A Curvilinear Relationship between Task Conflict and Innovation in Teams', *Journal of Management*, 32, 2006, 82–107.
48. Somech, Desivilya and Lidogoster, 'Team Conflict Management and Team Effectiveness: The Effects of Task Interdependence and Team Identification'.

49. Hempel, Zhang and Tjosvold, 'Conflict Management between and within Teams for Trusting Relationships and Performance in China'.
50. H.S. Desyvilya, A. Somech and H. Lidgoster, 'Innovation and Conflict Management in Work Teams: The Effects of Team Identification and Task and Relationship Conflict', *Negotiation and Conflict Management Research*, 3, 2010, 28–48.
51. Tajfel and J. C. Turner, 'The Social Identity Theory of Intergroup Behavior'.
52. Behfar, Kern and Brett, 'Managing Challenges in Multicultural Teams'.
53. F. A. Mael and B. E. Ashforth, 'Loyal from Day One: Biodata, Organizational Identification, and Turnover among Newcomers', *Personnel Psychology*, 48, 1995, 309–33.
54. P. C. Earley and E. Mosakowsky, 'Creating Hybrid Team Cultures: An Empirical Test of Transnational Team Functioning', *Academy of Management Journal*, 43(1), 2000, 26–49; Glinow, Shapiro and Brett, 'Can We Talk, and Should We? Managing Emotional Conflict in Multicultural Teams'.
55. G. S. van der Vegt and J. S. Bunderson, 'Learning and Performance in Multidisciplinary Teams: The Importance of Collective Team Identification', *Academy of Management Journal*, 48(3), 2005, 532–47.
56. S. Lembke and M. G. Wilson, 'Putting the "Team" into Teamwork: Alternative Theoretical Contributions for Contemporary Management Practice', *Human Relations*, 51, 1998, 927–44.
57. Ibid.
58. T. R. Tyler and S. Blader, 'Cooperation in Groups: Procedural Justice, Social Identity, and Behavioral Engagement', Psychology Press, Philadelphia, 2000.
59. Somech, Desivilya and Lidogoster, 'Team Conflict Management and Team Effectiveness: The Effects of Task Interdependence and Team Identification'.
60. M. Mortensen and P. Hinds, 'Conflict and Shared Identity in Geographically Distributed Teams', *International Journal of Conflict Management*, 12, 2001, 212–38.
61. G. Han and P. D. Harms, 'Team Identification, Trust, and Conflict: A Mediation Model', *International Journal of Conflict Management*, 21, 2010, 20–43.
62. N. Paulsen, D. Maldonado, V. J. Callan and O. B. Ayoko, 'Charismatic Leadership, Change and Innovation in R&D

Organisation', *Journal of Organizational Change Management*, 225, 2009, 511–23; Desyvilya, Somech and Lidgoster, 'Innovation and Conflict Management in Work Teams: The Effects of Team Identification and Task and Relationship Conflict'.

63. Paulsen, Maldonado, Callan and Ayoko, 'Charismatic Leadership, Change and Innovation in R&D Organisation'.
64. Desyvilya, Somech and Lidgoster, 'Innovation and Conflict Management in Work Teams: The Effects of Team Identification and Task and Relationship Conflict'.
65. J. R. Hackman and R. E. Walton, 'Leading Groups in Organizations' in *Designing Effective Work Groups* (ed. P. S. Goodman), Jossey-Bass, San Francisco, 1986, 72–119.
66. J. R. Hackman and R. Wageman, 'Foster Team Effectiveness by Fulfilling Key Leadership Functions', in *Handbook of Principles of Organizational Behavior* (ed. E. A. Locke), Wiley-Blackwell, New York, 2009, 275–93.
67. Ibid.
68. Hackman and Walton, 'Leading Groups in Organizations'.
69. M. A. Hogg, 'A Social Identity Theory of Leadership'. *Personality and Social Psychology Review*, 5, 2001, 184–200.
70. M. A. Hogg and D. van Knippenberg, 'Social Identity and Leadership Processes in Groups' in Advances in Experimental Social Psychology (ed. M. P. Zanna), 35, San Diego, Academic Press, 2003, 1–52.
71. Paulsen, Maldonado, Callan and Ayoko, 'Charismatic Leadership, Change and Innovation in R&D Organisation'.
72. V. Benet-Martinez, F. Lee and J. Leu, 'Biculturalism and Cognitive Complexity: Expertise in Cultural Representations', *Journal of Cross-Cultural Psychology*, 37(4), 2006, 386–407; C. T. Tadmor, P. E. Tetlock and K. Peng, 'Acculturation Strategies and Integrative Complexity: The Cognitive Implications of Biculturalism', *Journal of Cross-Cultural Psychology*, 40(1), 2009, 105–39.
73. C. Wan, C.-Y. Chiu, S. Peng and K.-P. Tam, 'Measuring Cultures through Inter-subjective Cultural Norms: Implications for Predicting Relative Identification with Two or More Cultures', *Journal of Cross-Cultural Psychology*, 38(2), 2007, 213–26.
74. S. J. Schwartz, M. J. Montgomery and E. Briones, 'The Role of Identity in Acculturation among Immigrant People: Theoretical Propositions, Empirical Questions, and Applied Recommendations', *Human Development*, 49(1), 2006, 1–30.

75. Tajfel and J. C. Turner, 'The Social Identity Theory of Intergroup Behavior'.
76. Ancona and Caldwell, 'Demography and Design: Predictors of New Product Team Performance'; Blau, *Inequality and Heterogeneity : A Primitive Theory of Social Structure*; Cox, *Cultural Diversity in Organizations: Theory, Research, and Practice.*

References

T. M. Amabile, 'The Social Psychology of Creativity', Springer-Verlag, New York, 1983.

J. Pfeffer, 'Organizational Demography', in *Research in Organizational Behavior* (eds L. L. Cummings and B. M. Staw), JAI Press, Greenwich, 1983, 229–357.

10

Strategic and Tactical Dimensions of Technology Standards Battles: Sony and Blu-Ray

C. Jayachandran and Ram Subramanian

Sony's experience with the Betamax in the 1970s, and the Blu-Ray more recently, provides a good opportunity to examine how technology standard battles are played out. The paper looks at the strategic and tactical decisions that firms make that determine the outcome of such battles. Using archival data and the extant literature on standards, the paper profiles the two battles to provide a contrasting look at Sony's decisions.

Standards are an integral part of the milieu in technology-based industries. Competing standards exist in many technology products. Examples of competing standards include the Mac versus the Windows operating systems, and the Xbox, Nintendo Wii, and the Playstation consoles in the video gaming industry. There are two critical elements confronting companies in their quest for profit from technology standards. The first has to do with innovation — in short, it requires the company to set the standard via new technological breakthroughs. The second is to appropriate value from the standard that the company creates. Value appropriation of technology standards requires the bandwagon effect to play out at an optimum level. Researchers have pointed out the impact of network externalities in technology-based industries. If the pioneering company succeeds in locking-in its technology, it is not only likely to earn above normal rents for a long time, it can also create entry barriers for other technology providers. Microsoft's near complete monopoly of the market for computer operating systems is a good example of the benefits of the bandwagon effect and of lock-in, and the benefits of network externalities.

Sony has been involved in two high-profile standards wars. In the 1960s, it lost a much publicized battle when its Betamax video format was supplanted in the market by the Matsushita-led Video Home System (VHS) format. However, Sony redeemed itself in 2008, when it prevailed with its Blu-Ray technology that beat out the competing High Definition–Digital Video Disc (HD-DVD) format spearheaded by Toshiba. Sony's contrasting experience provides a good lens to look at the strategic and tactical dimensions of technology standard battles.

The objective of this paper is to examine how organizations face significant choices in both the strategic and tactical aspects of competing in standards wars, and how many of these decisions have significant trade-offs in the value creation and value appropriation dimensions.

Theoretical Background

Three concepts underpin technology standards battles: the lock-in effect, network externalities, and winner-take-all (WTA) markets. As companies attempt to dominate the standard in incompatible technologies, they make strategic decisions to impose switching costs on customers, increase the benefit to both existing and potential customers in joining their technology network, and make competing technologies unprofitable.

As Jonathan Koch[1] argues in his seminal paper, strategic lock-in, when successful, imposes a switching cost to customers. When a customer is 'locked-in' to a product embodying a certain technology, the customer has chosen to invest considerable resources in that technology which may be incompatible with other competing technologies. This 'lock-in' means that both economical and non-economical (learning, wastage of time) costs implore the user to continue using the current product, and not switch to a competing product. An everyday example serves to illustrate this concept. When a large institution chooses to use a certain email software, it incurs costs to buy the software, train users and consultants to use the software, create help manuals to troubleshoot, and invests in several such expenditures. If the same organization is confronted with the need to switch to a different email software, it faces both economic (cost of the new software, training costs, etc.) as well as non-economic (user reluctance to learning a new software) costs, and hence would make a decision based upon switching costs. This gives an advantage to the incumbent product, which is at the heart of any technology product involved in a standards war.

The concept of network externalities came to the front during Microsoft's battle with the United States Justice Department in what came to be dubbed as the 'browser wars'. The Justice Department argued that by bundling Internet Explorer (IE) to Windows (the operating system software that drives a personal computer), Microsoft successfully created a network. This gave a significant advantage to Microsoft because a potential user could benefit by joining a large network (a large network meant more compatible products offered to users as compared to a product with a small network which may not have the scale to encourage producers of compatible products) and, in turn, increase the attractiveness of the network to current users. Economides,[2] Liebowitz and Margolis,[3] and Evans, Hagiu, and Schmalensee[4] examined the barriers to entry that the network effect creates for potential entrants to a market. Firms involved in technology standards battles yearn to benefit from the network effect as it, in effect, creates a 'virtuous cycle'. More recently, the network of iPod users helped create an iPod 'ecosystem' that helped cement Apple's dominance in mobile music devices.

Eisenmann[5] pioneered the concept of the 'winner-takes-all' (WTA) markets in technology-based industries. He argued that the lock-in and network effects create a bandwagon benefit to certain technologies, so much so, that it becomes extremely difficult or nearly impossible for competing technologies to survive. He dubbed such markets as WTA markets because such markets, in effect, become a monopoly that favors the winning standard. Thus, Google's nearly 80 percent market share in the search market, Microsoft's 90 percent plus market share in operating system software, and Apple's dominant position in music devices imply that it is unprofitable to be the second or the third player in such industries. The second place and third place firms in such markets would find themselves in a vicious cycle, because producers of complementors, such as application software, would be discouraged from making them for lack of a critical mass to be profitable. Simply put, in WTA markets, the winning standard benefits from a virtuous cycle while the 'losing' standards become victims of a vicious cycle.

The strategic and tactical decisions taken by a firm involved in a standards battle are aimed at benefitting from the effects of lock-in and network externalities, so that they end up as a WTA market. What makes this complicated is the fact that these decisions typically involve trade-offs between profit appropriation and the creation of a large user base. In other words, the decisions that such a firm makes have to involve examining two typically divergent consequences: holding on to their share of the profits on the one hand, and on the other, creating an ecosystem that encourages users to form a large network. Sony's experience with the Betamax in the 1970s and its more recent experience with the Blu-Ray technology, bring to the forefront the importance of paying careful attention to this trade-off.

Sony and Betamax

Sony launched its Betamax home movie player in 1975, a full year before the Matsushita-led consortium launched the competing Video Home System (VHS) product. The two systems were incompatible because of differences in cassette construction and the tape-drive mechanism.[6] In other words, users had to buy movies that played only on one system and not the other. The Betamax was priced at a USD 60 premium over the VHS, because it believed that its system offered better picture quality. In spite of its first-mover advantage and its putative picture quality, the market share advantage tipped in favor of the VHS in as early as

1978 — a mere two years after VHS's launch in 1976. The table below lists the yearly market share of Betamax and VHS over the 1978–1986 time periods.

Table 10.1 Market Share in Percentage

YEAR	*VHS*	*BETAMAX*
1978	56.8	43.2
1979	55.0	45.0
1980	60.0	40.0
1981	72.4	27.6
1982	74.4	25.6
1983	81.9	18.1
1984	82.2	17.8
1985	87.8	12.2
1986	91.4	8.6

Source: Ohashi, 2003

It was in 1983 that Zenith, a leader in branded televisions, made the decision to abandon the Betamax format and become an exclusive supplier of VHS-compatible products. In the later stages of this battle, Sony used the positioning strategy of a higher priced (higher compared to its early differential with VHS) niche product, but it did not help the swift decline of its market share.

Sony and Blu-Ray

More recently, a technology-standards battle developed, pitting Sony and its Blu-Ray technology with the Toshiba-led HD-DVD technology. It was once again in the home movie product area, except that technology in this space had shifted to the digital from the analog arena.

In 2003, Sony announced that it was developing a new technology, called Blu-Ray, to offer state-of-the-art digital viewing of movies for the home user. One year later, in 2004, a consortium led by Toshiba announced the development of the competing HD-DVD technology for the same market. In essence, this standards battle began in 2004. It ended in 2008 when Toshiba announced that it was withdrawing the HD-DVD product from the market, thereby conceding the market to Sony's Blu-Ray. Compared to the 11-year battle with the Betamax and VHS (1975–1986), the Blu-Ray/HD-DVD battle was short-lived, probably indicating that the stakes and resources were much higher the second time around, and hence called for a quick decision whether to continue further investment or cut the losses early.

Strategic and Tactical Dimensions: A Comparison

Seifert and Vare[7] posit that every standards battle has a strategic and tactical dimension involved. When a firm is involved in a standards battle, it makes a series of decisions, some of whom have a strategic orientation, while others are more tactical in nature. Using the framework provided by Seifert and Vare,[8] we developed a model that captures the set of decisions involved in a standards battle.

The major trade-off in any technology-standards battle is between standard setting and value appropriation. A trade-off occurs because the two may be incompatible. Standard setting involves market acceptance and the consequent creation of a large installed user base.[9] This means pricing the product such that potential users are attracted to buy the product when faced with technology choices. Sony's pricing of the Betamax at a premium — a USD 60 price premium to begin with, that increased later on when Sony tried to position it as a niche product—indicated a choice that favored profit appropriation over market acceptance. A similar such scenario occurred in IBM's PC (personal computer) product in comparison to Apple's Mac. Value appropriation becomes paramount when the product technology involves creating an ecosystem. Such an ecosystem involves decisions regarding forming strategic alliances, licensing the technology to other manufacturers, product pricing, backwards compatibility (compatibility with a previous generation product), and finally, social networks. Strategic and tactical decisions in the standard setting arena are essentially oriented to creating a large installed base to benefit from network externalities.

Figure 10.1 A Theoretical Model of the Strategic and Tactical Dimensions of Technology-Standards Battles

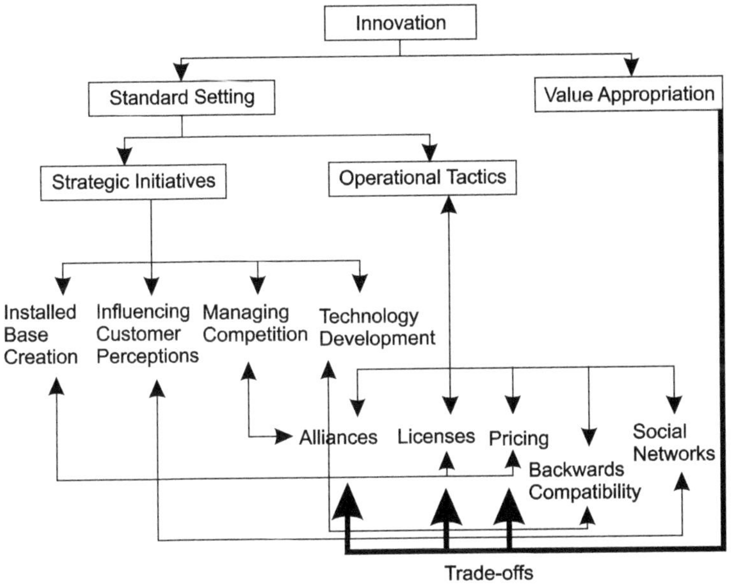

Source: Adapted from Seifert and Vare[10]

It is clear that Sony did not fully consider the trade-offs involved in a technology standards battle in its Betamax launch. It launched the Betamax first and then looked for consortium partners, thereby delaying the benefits of a concerted attack to increase the user base.[11] It used what is referred to as the 'closed garden loop' in not allowing its allies to put their own brand name on the product. All Betamax products were sold under the Betamax brand name regardless of who the manufacturer was, unlike the case of the VHS where the manufacturer's brand name was prominent. As Ohashi[12] points out, the VHS consortium looked for network externalities right from the beginning, even if it meant that Matsushita—the current incarnation of what was then the Victor Company of Japan—had to settle for a reduced profit share. Since a movie player's value increased with the availability of movies for that format (what is referred to as a 'complementor'), a large user base encouraged movie studios to invest in making movies for that format that created a virtuous cycle in favor of the VHS consortium. As Table 10.1 points out, while the market was fairly even up until 1979 (when Betamax's share was 45 percent to VHS's 55 percent), it began to tip from 1980 onwards because of the wide availability of comparatively cheaper movies for the VHS format. The 'virtuous cycle' continued for VHS as its market share gap with the Betamax increased to a dominant 91.4 percent in 1986, effectively ending the battle. The VHS consortium, in contrast to Sony, recognized the trade-offs involved and actively sought licensees and alliance partners to increase the user base. These were underscored by the strategic decisions taken by the consortium.

It appears that Sony recognized the trade-offs in technology battles clearly in its second effort at such battles, when it developed the Blu-Ray format. It gave up significantly on the value appropriation frontier both by its pricing in comparison to the HD-DVD format and by its encouragement of consortium partners. Sony sought partners both to produce the hardware (the player) as well as in providing complementors (the movie studios and players such as Netflix and Blockbuster). Sony also made the decision to include the Blu-Ray player in its PlayStation product, thereby benefitting from an existing user base for that product. It obtained exclusive contracts with Disney, 20th Century Fox, and Lionsgate (in addition to its own movie studio), while obtaining non-exclusive rights with Warner Bros., and MGM.[13] Compared to this, Toshiba's HD-DVD consortium had exclusivity with Paramount, Universal, and the much smaller Dreamworks. In terms of movie market share, Sony's exclusive partners accounted for 37.9 percent of the market versus 24.4 percent for the rival consortium.[14] In particular, Warner Bros. decision to also support Blu-Ray (after its initial decision to support HD-DVD exclusively) in late 2006/early 2007, appears to have tipped the battle in favor of Blu-Ray. The Warner decision gave the Blu-Ray consortium partners the impetus to step up the release of movies for this format. As Christ and Slowak[15] point out, the release of movies for both formats was neck-and-neck until this point, after which movies for the Blu-Ray increased in geometric proportion. Sony's cultivation of the complementors in its ecosystem helped pave the way for the eventual dominance of its format, something it either chose not to do with Betamax or chose it too late.

Discussion

Technology-standards battles in most industries are high-stake wars that eventually result in WTA markets. To win a technology standards battle, firms have to consider the myriad trade-offs that occur between market acceptance and profit appropriation. From a short-term point-of-view, a firm may be giving up profits. But, if in return, it obtains market acceptance, it is likely to come out as the winner in such contests. These decisions involving trade-offs occur within both the strategic and the tactical dimensions of the battle.

Sony's experience with two such battles—one where it came out as the loser and the other where it was the winner — illustrates the consequence of key choices made in both the strategic and the tactical aspects of technology standards. While some may claim that Sony's victory in Blu-Ray was pyrrhic (because it may end up making very little money off it), the virtuous cycle that it successfully created in the latter battle allowed it to become the dominant player in this format, which helps the movie arm of this conglomerate as well as its Playstation unit.

Three key aspects of technology standards, lock-in, network externalities, and WTA markets, combined with the need to effectively create a virtuous cycle, characterize players who have succeeded in such wars. From a strategic point of view, this means looking at such battles from the long-term perspective. Both Apple in its Mac, and Sony in its Betamax, looked at immediate gains as the precursor to long-term profitability. What they failed to do was to create an ecosystem of consortium partners, who may take away a part of the focal firm's profits in the short-run, but allow the focal firm to outflank the competition in the long-run. An effective technology battle is fought by a consortium rather than by an individual firm. The creation of a consortium means forgoing some of current profits for future gains. In both the cases of the Betamax and Blu-Ray, the tipping point came from the action of a consortium partner. Zenith's decision to focus exclusively on the VHS format was the final straw that broke Betamax. Similarly, Warner Bros' decision to support the Blu-Ray format also, allowed Sony to overtake the HD-DVD format. It is important to remember that neither action (Zenith's and Warner Bros') was the action of the focal firm.

In the technology domain, we are currently witnessing standards battles in mobile telecommunications (Apple versus Google, for example), and mobile music (Apple versus others). The race in these cases is to create a large and effective ecosystem consisting of erstwhile competitors as well as players in the complementors market. As the focal firms in these battles gear up for the fight, it may behoove them to consider the trade-offs that they need to confront in both the strategic and the tactical dimensions of standards wars.

Notes

1. J. Koch, 'Inscribed Strategies: Exploring the Organizational Nature of Strategic Lock-in', *Organization Studies*, 32(3), 2011, 337–58.
2. N. Economides, 'The Economics of Networks', *International Journal of Industrial Economics*, 14, 1996, 673–99.
3. S. Leibovitz and S. Margolis, 'Network Externality: An Uncommon Tragedy', *Journal of Economic Perspectives*, 2, 1994, 133–50.
4. D. Evans, A. Hagiu and R. Schmalensee, 'How Software Platforms Drive Innovation and Transform Industries', MIT Press, Boston, 2006.
5. T. Eisenmann, 'Note on Racing to Acquire Customers', *Harvard Business Publishing Note*, 2007.
6. J. P. Christ and A. P. Slowak, 'Why Blu-Ray vs. HD-DVD is not VHS vs. Betamax: The Co-evolution of Standard-setting Consortia', Working Paper 29, Stuttgart, 2009.
7. R.W. Seifert and A.X. Vare, 'Fighting Technology Standards Competitions: Adapting to New Rules', *Perspectives for Managers*, IMD, 158, 2008.
8. Seifert and Vare, 'Fighting Technology Standards Competitions: Adapting to New Rules'.

9. R. Grant, 'Contemporary Strategy Analysis', Wiley, New York, 2010.
10. Seifert and Vare, 'Fighting Technology Standards Competitions: Adapting to New Rules'.
11. M. A. Cusumano, Y. Mylonadis and R. S. Rosenbloom, 'Strategic Maneuvering and Mass Market Dynamics: The Triumph of VHS over Beta', *Business History Review*, 66(1), 1992, 51–94.
12. H. Ohashi, 'The Role of Network Effects in the U.S. VCR Market, 1978–1986', *Journal of Economic and Management Strategy*, 12(4), 2003, 447–94.
13. Christ and Slowak, 'Why Blu-Ray vs. HD-DVD is not VHS vs. Betamax: The Co-evolution of Standard-setting Consortia'.
14. Ibid.
15. Ibid.

References

D. F. Spulber, 'Unlocking Technology: Antitrust and Innovation', *The Journal of Competition Law and Economics*, 4(4), 2008, 915–66.

11

Challenges in the Development of Technological Capabilities: Business Models and Entrepreneurial Culture in Emerging Economies

Ricardo Arechavala-Vargas, Berta E. Madrigal-Torres, Bernardo Jaen-Jimenez and Lorena G. Verde-Flota

Small and medium enterprises (SMEs) in emerging economies tend not to use technological innovation as a competitive strategy, but rather use it passively to assimilate technology as embedded in capital goods. However, as these economies become increasingly subject to international competitive pressures, the need for improving technological capabilities increases in many sectors, while many conditions in the economic environment are still not conducive to the development of technological capabilities in small firms.

We present here results from a study that uses a database of approximately 400 SMEs in Mexico, and a set of 35 in-depth interviews dealing with the ways in which entrepreneurs and owner-managers perceive the need for greater technological capabilities, and the process through which they develop those capabilities. Results helped us identify some of the key managerial and entrepreneurial characteristics associated with the development of technological innovation capabilities in the firm. We also identified characteristics in the business environment that constrain or enhance their chances of success in this endeavor, and discuss policy implications for regional and federal programs that aim to support SMEs.

Business Strategy and Technological Capabilities in Emerging Economies

An important fact in describing and understanding the role that technology and innovation play in emerging economies is that firms in them have had to compete in their own domestic markets with multinational corporations, and with more mature firms that have been evolving for decades before them. SMEs in emergent economies have had to compete against firms that are way ahead of them in learning and developing business capabilities. Firms in developing countries tend to adapt knowledge that has been developed in industrialized countries to their industrial sectors' absorptive capacity and local market demand,[1] or to simply adopt technologies as incorporated in capital goods. In some instances, they even use obsolete machinery that has been discarded from more competitive economies.

Drawing on technology acquired abroad or assimilated from foreign-based firms, local firms imitate, copy, or generate products that are technologically similar to existing ones in developed economies.[2] Firms adapt such knowledge to local needs, a process which usually does not require cooperation with organizations like universities or banks, or among firms themselves.[3]

A useful conceptual framework in trying to understand constraints that small firms face when attempting to adapt to new competitive environments, can be built on the basis of theoretical advances in the area of business models and business model dynamics.[4] As firms make choices in order to create and capture value, they respond to conditions in their environment that represent opportunities or threats. Whether these decisions correspond to an explicit design, or whether they are reactive responses to unforeseen situations, they configure business models that define their strategies in creating and capturing value, the operational architecture they will use to achieve this, and their interaction patterns with other organizations in their environment.

Initially, almost every firm will have a period of experimentation, as they learn along the way which strategies and operational configurations are more conducive to their survival and growth. However, as long as they are successful in this task, they will tend to stabilize over time, switching their skills from experimentation to the exploitation of fruitful operational arrangements.

Once their choices give rise to a specific business model, their strategic and tactical choices are constrained: they cannot easily take advantage of, or in many instances even realize, opportunities that their business model is not well equipped to handle.[5] As technology and innovation is not usually part of their business model in emerging economies, entrepreneurs and managers in small firms will tend not to consider options in that arena when faced with competitive pressures. In many instances, particularly in the Mexican case, import substitution policies precisely created a context, in which technology was something to be acquired as part of capital goods, instead of being created in-house. However, as markets have rapidly opened in recent years, such firms have faced competitive pressures and they are ill-equipped to handle them. After decades of operating under business models that did not require endogenous innovation and technology, they have had enormous difficulty in realizing the need for them, and even when they do realize this need, they have had little resources and capital to develop the required capabilities.

Firms in less-developed economies, have also had to compete on the basis of business ecosystems that provide them with lower levels and quality of support from other organizations.[6] Thus, firms in industrialized nations that come to operate in countries with less developed economies, face weak competition, and are able to easily establish technological superiority and market dominance, while endogenous firms face lack of high-level human resources and technological infrastructure to compete, thus being relegated to compete on the basis of cheap unskilled labor in their products and production processes, and therefore to low value-added markets.[7]

In this context, firms created in emerging economies tend to seek lower value-added markets in order to survive, and to acquire technology developed abroad as embodied in capital goods, which are usually obsolete in the countries where they were produced. Financial capital tends to be scarce, and concentrated in few families, so firms are set in a context where survival depends on lower prices, rather than on differentiation and innovation. As has happened in many developing countries (such as Mexico, Brazil and Chile, for example), government policies have tended to be protectionist, resorting, at least for some periods, to import substitution strategies. In very few cases, such as Korea, China and other newly industrialized countries, they have aggressive policies aimed at developing endogenous technological capabilities.

In this way, firms in emerging economies have had little resources, little government support, and weak business environments, on the basis of which they cannot compete internationally or even defend their domestic markets. Technological innovation has therefore been absent from their business practices in most cases, and business education has only been rudimentary, and based on trial and error survival tactics. Moreover, previous studies have documented the fact that Mexican firms tend to exhibit a low level of adaptability to changing circumstances, and that cohesiveness of top management teams around founders' centrality in decision making, tends to become a strategic liability for them.[8]

Thus, small firms in Mexico, as very likely in the rest of Latin America, tend to remain rigidly fixated on old business models and practices, even in the light of increasing domestic and international competition. Few of them have embarked in processes that aim to improve their competitiveness. Once the initial stages of experimentation and survival give way to a relatively stable operations, they tend to give up searches for better business models and more competitive operating practices, and choose to exploit old and well-known routines, as opposed to learning new behaviours.[9]

More recently, however, a growing segment of local firms have been able to compete out of low value-added markets, and into more profitable ones. Whether they have learned along the way—usually across generations—or whether they are founded by entrepreneurs with international-level education and outlook, these firms are beginning to systematically use technological innovation to successfully compete in both domestic and international markets.[10]

However, in view of the fact that many small firms in western Mexico have shown a lack of capability for change in their strategies and business models, there are several questions that arise when trying to understand those factors that may influence their ability to learn, to mature towards better practices, and be more competitive. How do firms shape their business models, and how do these business models shape the search space in which they seek tactical and operational solutions to the problems they face? Do firms evolve, at least to a significant extent, as a function of firm age? Do they, for example, initiate their existence with very rudimentary business practices and operating systems, and later evolve them into more sophisticated versions, or are their business models and routines imprinted rigidly into their structures and culture, once they are able to survive in the market? Is their ability to evolve into higher order capabilities a function of the entrepreneurial profile of their founder? Do such capabilities evolve as business goals grow more ambitious over time?

Thus, the goal of this study is to search for variables that may account for a firm's propensity to evolve towards international-level practices and systems, in order to jump out of the spiraling trap that falling back into low value-added markets represents. An understanding of these processes should be useful in improving the overall competitiveness of national economies by guiding policy-making; it should also be relevant in pulling more traditional and less competitive industries forward, in order to avoid the gap between traditional and more dynamic sectors within an economy.

Method

As part of our group's consulting and research activities, case studies are systematically conducted, and subjected to a diagnostic instrument designed to evaluate the firms' level of development, with respect to a set of more than 70 variables. Variables are defined in each case as Guttman scales with five levels each.

Variables are grouped by functional area—management, production, commercialization, finance, human resources and organization—and three groups of variables that cut across functional areas of business culture, technological capabilities, and exports capabilities. Case-study coding for all of these variables is carefully tailored to ensure inter-observer reliability: each case study is coded by three independent groups, and in those cases where significant discrepancies are found, they are excluded from the database.

All of the variables are defined in terms of actual practices and behaviors observed in firms, rather than on the basis of normative managerial recommendations, as is usually the case with operational audits. Each variable's lower values reflect those practices used by firms with less experience, technical level and capabilities, while higher levels describe more sophisticated and powerful operating procedures, approaching world class firms' behavior.

Variables selected for this analysis correspond to theoretically based concepts. Entrepreneurial orientation, for example, is a construct that represents management's proactive, goals and strategic outlook,[11] which has also been grounded on empirical evidence.[12] Technological capabilities, as relevant to firms in emerging economies, follow Bell and Pavitt's approach,[13] which has also been extensively documented for studies of Latin American firms. The database used has been coded with these constructs in mind. For this analysis, a database of 363 small- and medium-sized firms in Western Mexico were used. Mean age of the firms is 25.15 years, and mean size is 61.8 employees. Firms included in the database come from different industrial and service sectors, as shown in Table 11.1. No attempt was made to obtain a stratified sample, since the main constraint in the country is usually the extreme secretiveness of owners and founders of firms, which leads them to refuse access to outsiders. In many instances, this access was obtained mainly on the basis of consulting activities that our group carries out regularly, in order to develop more competitive capabilities in them.

Table 11.1 Industry Sectors Included in the Sample

Industry Sector	*No.*
Agro	7
Food and Beverages	58
Crafts	9
Shoe	30
Garment	40
Construction	10
Transport	3
Commerce	37
Electronics	8
Printing and Graphics	5
Furniture	22
Metal mechanics	21
Plastics	23
Chemicals and pharmaceuticals	8
Services	82
	n = 363

Thirty five semi-structured interviews with higher management in firms (usually their owner/founder) were conducted in order to assess their accounts of how they developed specific business and technological capabilities, and how these capabilities relate to their business goals and strategies. Interviews were designed in order to qualitatively assess entrepreneurs' backgrounds, how they acquired their managerial and technical capabilities, and their business goals and strategies, with the goal of inductively identifying patterns in their origins as entrepreneurs, and in their business/managerial culture.

Interview transcripts were then coded with the aid of a content-analysis software (QSR NUD*IST), in order to ascertain the relationship of these goals and strategies with theoretically based categories, and in order to inductively build new categories that may be more adequate in describing small firm behavior in emerging economies.

Even though qualitative data is not easy to integrate as scientific evidence, it does help understand the ways in which quantitatively measured variables interact in specific cases, thus providing a better foundation for the comprehension of causal relationships that they quantitatively assess.[14]

Results

In order to quantitatively assess patterns in learning behaviors among firms that were studied, a subset of variables reflecting processes likely to influence a firm's proclivity to improve their operating systems and procedures toward more competitive levels was selected. Data shows that there are important differences among those variables studied, as predictors of overall maturity level (OVALLMAT) that a firm achieves, in terms of the set of variables coded in the database, excluding the values of the variables used as predictors. Table 11.2 summarizes correlation coefficients for the variables studied.

Table 11.2 Correlations among Business Culture Variables

	OVALLMAT	*FIRMAGE*	*GOALS*	*SIZE*	*ENTREPOR*
OVALLMAT	-				
FIRMAGE	0.338	-			
GOALS	0.561	0.126	-		
SIZE	0.308	0.331	0.193	-	
ENTREPOR	0.828	0.251	0.698	0.261	-
IDT	0.841	0.312	0.407	0.298	0.606

Because of the size of the sample, all of the variables selected for this study have a statistically significant ($p<0.01$) correlation with overall maturity level (OVALLMAT), which is our dependent variable. However, it is worthwhile to note that the highest correlations with this variable correspond to entrepreneurial orientation (ENTREPOR) and technological capabilities (TECHCAP), while the smallest correlations exist between firm age (FIRMAGE) and management's goals (GOALS) on the one hand, and firm size (SIZE) and GOALS on the other. At first glance, this can be interpreted as a relatively weak relationship between the size of the firm in terms of number of workers and its business goals. That is to say, the level of the firm's goals is only weakly correlated to its age or to its size, as measured by its number of workers. Contrary to what could be expected, the correlation between firm age and its overall maturity level is also among the lowest values in the table. That is to say, a firm's overall competitiveness is not merely a function of its age: firms do not necessarily continue to evolve into more advanced practices over time. Also, the relatively low value (0.338) of the correlation between a firm's size and its overall maturity level, seems to indicate that its competitiveness is not primarily a function of its growth. Although this correlation is significant—as are all of those included in the table simply because of the sample size— its value is smaller than others.

On the other hand, the significantly higher correlation between the overall competitiveness of the firm and its technological capabilities (TECHCAP), calls attention to the close relationship that exists between a firm's level of technological capabilities and its ability to evolve into world-class standards. Together with entrepreneurial orientation in the management team (ENTREPOR), technological capabilities in the firm are an important predictor of its competitiveness.

Interviewees' accounts of their entrepreneurial development show qualitative differences in their training and business experience, that provide insights into the nature of the processes, that account for the associations among the variables described above. More technologically dynamic firms, as compared to less flexible and more traditional firms for example, tend to be founded by technically trained entrepreneurs with experience in large transnational firms or with international-level education. This background provides them with both technical expertise and more competitive business and managerial skills. In what follows, some illustrative responses are reproduced:

Q: *How did you acquire the managerial skills necessary in founding and developing your firm?*

A: *First, because of my professional training, and second because of the opportunity I had to work for different major laboratories for many years, where I participated in the development of production processes, new products, marketing; and all of that helped me prepare to start my own business.*

A significant difference is also evident when comparing interviewees' accounts of the way in which they have built a successful firm. Such difference has to do with their predisposition and ability to interact with other organizations and sources of knowledge in order to solve technical problems:

Well, as far as technological capabilities, we rely on support from the (local public) university and from an external advisor, who helps us in developing synergistic formulations for our products . . . that enable us to use antibiotics that increase the range of effects on the animals and have good results. That helps us a lot . . .

. . . commercially we have opened distribution channels and tried to go directly to producers, we leverage trade associations . . . we also participate in national and local trade exhibitions and conferences . . . and try to reach decision makers, be they veterinarians, farm owners and the like . . .

Learning and knowledge management are important for technologically dynamic firms, as opposed to more traditional firms. In technologically active firms, usually proactive views in this aspect shape decision-making in most functional areas:

Our administrative manager is continually in training, be it on accounting and fiscal requirements and procedures, in human resource management, and so on . . . the purchasing manager has to deal with raw material imports, dealing with international and local suppliers and needs to participate in updating courses frequently, as well as our sales assistants and production supervisors.

The same is not true, however, for more traditional firms, where personal experience and knowledge of a trade have frequently been the basis of stability for the firm. Patterns in their responses reveal a lack of entrepreneurial orientation, as well as little predisposition towards collaboration with other organizations, and development of new skills and capabilities, which usually are the product of the firm's strategic space as defined by its founder:

Q: *How did you acquire the managerial skills necessary in founding and developing your firm?*

A: *Well, it is rather the founder's skills. He used to be a physician, but then went on to create the company. He is the one who knows the business. . . . in the technology arena we are obsolete. I know of new, bigger, faster machines that are now used in (printing industry), but we don't have them . . . there haven't been significant changes in the company in a long time . . . No, we do not collaborate with other firms in this or other industries . . . firms should be independent, printing firms should always be independent.*

Q: *How do you see your firm in five years? In 10 years?*

A: *I hope we do get to five years hence, I hope we can continue to operate the same way as we are now, at least the same as we are now . . . 10 years? 10 years is even more uncertain, just imagine . . . 2019 . . . if we can get over the present crisis I hope we are doing better.*

Thus, interviews show the ways in which variables of interest are related to each other in specific cases, even though the corresponding evidence can only be shown in an illustrative manner. In this last interview fragments, for example, it can be seen that firms with more elementary operating procedures tend to shy away from technology and from interaction with other organizations. This isolation tends, in its turn, to respond to those characteristics and values that founders imprint in the firms culture.[15]

Discussion

It seems surprising that the OVALLMAT of a firm is not primarily a function of its age or size, since it can be argued that firms that survive and grow over time would do so mostly because of their ability to adapt to changes in their environments, and to improve on their business models and practices.

The fact that correlation of overall maturity level with technological capabilities is greater than the correlation between overall maturity level and entrepreneurial orientation and this seems to imply that technological capabilities are a better indicator of overall maturity level for the firm, than its proactiveness and entrepreneurial orientation.

The fact that business culture variables, such as goals and entrepreneurial orientation, are less correlated than technological capabilities to overall maturity level indicates, that a firm's ability to evolve into world class practices and operating systems, is more strongly manifested in those organizations, in which technology has been adopted as a tool in competitive strategy. That is to say, when incorporated into their business models, technology is likely to represent a source of pressure towards further modifying the firm's architecture and operating systems. Technological change represents increasing demands and opportunities for change in the firm's strategic and tactical choices, thus providing a continuous influence in the firm's ability to adapt to external circumstances. This would make sense, as technologies tend to have a short life-cycle, and obsolescence is always a looming danger, but also because technological capabilities represent opportunities for further enhance the firm's ability to create and capture value. Besides that, technological change tends to imply organizational change, at least in the long run. Both processes would tend to push the firm towards continuous experimentation and exploration in its business practices.

It should be evident, however, that a directionality in the causal relationship between technological capabilities and overall maturity level in the firm cannot be established, since overall competitiveness would push the firm towards searching for more technological capabilities as it operates in more competitive environments, but the process in the opposite direction seems concomitant as well.

These previous arguments are reinforced by the fact that interviewees provide credible and consistent qualitative accounts of the way, in which developing technological capabilities is associated with a different mindset in the case of more technologically dynamic firms. Although not statistically oriented at this stage in our research, this analysis provides evidence of differences in patterns and backgrounds in more traditional firms, as opposed to those that have developed more significant technological capabilities.

The contribution that this study aims to make to the scientific literature, however, rests more on the fact that the correlations documented here call attention to processes that have still not been sufficiently documented in previous studies. More specifically, we aim to contribute to the understanding of those processes that trigger—or fail to do so—firm dynamism and learning in less developed countries. This understanding is all the more important because governments need it in order to devise policies that may improve these processes for the firms.

The fact that the level of technological capabilities in firms is such a strong predictor of their competitive level, for example, should point in the direction of policies and programs that increase incentives for firms, to improve their efforts to use technology and innovation as part of their survival and growth strategies. That, by itself, should signal to policy makers that the country, for example, should invest more in science, technology and innovation.

Such knowledge should also be important for practitioners such as business consultants, and for entrepreneurs, and higher management in firms struggling to gain competitiveness in emerging economies. Perhaps, for many of them, it would turn out to be too late to attempt to reorient their strategies towards a more technologically solid approach. Research and Development (R&D) is a most expensive, though profitable endeavour, and most of them are running out of investment resources, but the alternative is even less attractive.

Notes

1. J.A. Mathews, 'National Systems of Economic Learning: The Case of Technology Diffusion Management in East Asia', *International Journal of Technology Management*, 22(5/6), 2001, 455–79; E. B. Viotti, 'National Learning Systems. A New Approach on Technical Change in Late Industrializing Economies and Evidences from the Cases of Brazil and South Korea', *Technological Forecasting & Social Change*, 69, 2002, 653–80; J. Katz, 'Structural Reforms and Technological Behaviour: The Sources and Nature of Technological Change in Latin America in the 1990s', *Research Policy*, 30(1), 2001, 1–19.

2. Katz, 'Structural Reforms and Technological Behaviour: The Sources and Nature of Technological Change in Latin America in the 1990s'; P.S. Aulakh, M. Kotabe and H. Teegen, 'Export Strategies and Performance of Firms from Emerging Economies: Evidence from Brazil, Chile, and Mexico', *The Academy of Management Journal*, 43(3), 2000, 342–61.
3. H. Kroll and I. Liefner, 'Spin-off Enterprises as a Means of Technology Commercialisation in a Transforming Economy—Evidence from Three Universities in China', *Technovation*, 5, 2008.
4. M. Morris, M. Schindehutte and J. Allen, 'The Entrepreneur's Business Model: Toward a Unified Perspective', *Journal of Business Research*, 58(6), 2005, 726–35; A. Osterwalder, Y. Pigneur and C. L. Tucci, 'Clarifying Business Models: Origins, Present, and Future of the Concept', *Communications of AIS*, 15, May 2005, 1–40; R. Casadesus-Masanell and J. E. Ricart, 'From Strategy to Business Models and onto Tactics', *Long Range Planning*, 43(2–3), April 2010, 195–215.
5. Casadesus-Masanell and Ricart, 'From Strategy to Business Models and onto Tactics'.
6. M. Bell and K. Pavitt, 'Technological Accumulation and Industrial Growth: Contrasts between Developed and Developing Countries', *Industrial and Corporate Change*, 2(1), 1993, 157–210; S. Iammarino, R. Padilla-Pérez and N. von Tunzelmann, 'Technological Capabilities and Global-Local Interactions: The Electronics Industry in Two Mexican Regions', *World Development*, 36(10), 2008, 1980–2003.
7. Katz, 'Structural Reforms and Technological Behaviour: The Sources and Nature of Technological Change in Latin America in the 1990s'; Kroll and Liefner, 'Spin-off Enterprises as a Means of Technology Commercialisation in a Transforming Economy—Evidence from Three Universities in China'.
8. N. Athanassiou, W. F. Crittenden, L. M. Kelly and P. Marquez, 'Founder Centrality Effects on the Mexican Family Firm's Top Management Group: Firm Culture, Strategic Vision and Goals, and Firm Performance', *Journal of World Business*, 37(2), 2002, 139–50; C. R. Greer and G. K. Stephens, 'Escalation of Commitment: A Comparison of Differences between Mexican and U.S. Decision-Makers', *Journal of Management*, 27(1), 2001, 51–78.
9. J. G. March, 'Exploration and Exploitation in Organizational Learning', *Organization Science*, 2(1), 1991, 71–87.
10. C. Perez and L. Soete, 'Catching Up in Technology: Entry

Barriers and Windows of Opportunity', in *Technical Change and Economic Theory*, Pinter Publisher, New York, 1988; Iammarino, Padilla-Pérez and von Tunzelmann, 'Technological Capabilities and Global-Local Interactions: The Electronics Industry in Two Mexican Regions'; M. Fontes and R. Coombs, 'Contribution of New Technology-based Firms to the Strengthening of Technological Capabilities in Intermediate Economies', *Research Policy*, 30(1), 2001, 79–97.

11. G.T. Lumpkin and G. G. Dess, 'Clarifying the Entrepreneurial Orientation Construct and Linking It to Performance', *The Academy of Management Review*, 21(1), 1996, 135–72.
12. C.L. Wang, 'Entrepreneurial Orientation, Learning Orientation, and Firm Performance', *Entrepreneurship Theory and Practice*, 32(4), 2008, 635–57.
13. M. Bell and K. Pavitt, 'The Development of Technological Capabilities', in *International Competitiveness: Interaction of the Public and the Private Sectors* (ed. I. Haque), Washington DC, World Bank, 1995, 69–101; Bell and Pavitt, 'Technological Accumulation and Industrial Growth: Contrasts between Developed and Developing Countries'.
14. A.S. Lee, 'Integrating Positivist and Interpretive Approaches to Organizational Research', *Organization Science*, 2(4), 1991, 342–65.
15. O. Jones, A. Macpherson and R. Thorpe, 'Learning in Owner-managed Small Firms: Mediating Artefacts and Strategic Space', *Entrepreneurship & Regional Development: An International Journal*, 22(7), 2010, 649; E. Schein, *Organizational Culture and Leadership*, San Francisco, Jossey-Bass, 1985.

About the Editors

Ram Subramanian is Professor of Management, School of Business, Montclair State University. He received his PhD from the University of North Texas. He has published in a variety of journals including *Journal of Management*, *Journal of Business Research*, *Management International Review* and *Case Research Journal*. He serves on the editorial board of *The CASE Journal*.

Martin Rahe is Director of the Research Department and Professor of Economics and Business Policy, EADA Business School (Escuela de Alta Dirección y Administración), Barcelona. He holds a PhD in Economics from the University of Bochum (Germany) and has previously worked as a project manager at the Institute of Applied Innovation Research, where he handled partnerships between German and Russian companies. His research focus is on organizational behavior and on behavioral economics. He has published his research in a number of journal articles and books and is on the editorial board of several academic journals. He is a founding member of the Society for Global Business and Economic Development (SGBED).

Vishnuprasad Nagadevara is Adjunct Professor, Quantitative Methods and Information Systems Area, Indian Institute of Management, Bangalore. His current interests are data mining techniques and applications, application of operations research techniques, project management, project evaluation and information technology applications. He has carried out many research and consultancy projects for various organizations such as UNDP, WIPRO, POWERGRID, ETA (UAE), WiseScope (UAE), Fidelity, PSC, etc. He has published extensively in national and international journals.

C. Jayachandran is Professor of Marketing & International Business, School of Business, Montclair State University. His teaching, research and consulting interests are in international marketing, international business strategies of firms from emerging markets, business clusters, technology and innovation. He holds a PhD in Economics from the University of Madras. Some of his research papers have been published in a range of journals such as the *International Journal of Business Innovation & Research*, *Journal of Consumer Marketing*, *Economic and Political Weekly*, and *Economic Times,* among others. He is also the founding president of the Society for Global Business and Economic Development (SGBED).

Notes on Contributors

V. Richard Benjamins is Director of Business Intelligence at Telefonica Digital, the digital arm of the Telefonica Group. Before joining Telefonica, he co-funded and worked at iSOCO, a spin-off of the Spanish Institute of Artificial Intelligence. He has also worked in several universities over the world, published over 100 articles, and acquired, participated in and managed many international R&D projects, both in and outside Europe. He completed his PhD in Cognitive Science from the University of Amsterdam in 1993.

Pelin Bicen is an Assistant Professor of Marketing at the Sam and Irene Black School of Business, Penn State University, Erie, Pennsylvania. In her research, she investigates intra- and inter-organizational new product development activities of firms, resource limitations and innovation, and new product creativity. Her research has been published in the *Journal of Statistical Research, International Journal of Retail & Distribution Management, Journal of Marketing Education, Journal of Business Research, Journal of Business and Industrial Marketing*, among others. She has also presented her research in many national and international conferences and workshops.

Pankaj Chandra is Professor, Production and Operations Management and Former Director, Indian Institute of Management, Bangalore. He holds a B.Tech from Banaras Hindu University and a PhD from The Wharton School, University of Pennsylvania. He has taught at McGill University, Montreal; University of Geneva; The Wharton School; and IIM Ahmedabad, where he was also the Chairperson of the Doctoral Programme and of the Centre for Innovation, Incubation and Entrepreneurship. He has published extensively and serves on the editorial board of several international journals.

Maria V. Ciasullo is Associate Professor of Business Management, Salerno's University, Italy. She is a member of the body of PhD in Management & Information Technology at the University of Salerno's Department of Study and Business Research and holds a PhD in Business Management. Her research themes relate to strategic management and corporate governance of public and private organizations. Her papers have been published in national and international journals.

Jesus Contreras is Chief Operating Officer at playence. Denodo. He completed his PhD in Computer Science from the Technical University of Madrid, where he also collaborated as an Assistant Researcher in the Intelligence Systems Research Group. He holds an MBA degree and has also worked in several companies in leading positions. He has been the Innovation Director in iSOCEand Denodo. He has published various articles about innovation natural language processing and human-computer interaction.

Magdalena Bialic-Davendra has a Master of Arts (M.A.) in Economics from the University of Rzeszów, Poland, and PhD in Finance from Tomas Bata University in Zlín, Czech Republic. Her research profile includes being author and co-author of more than 25 scientific articles, a monograph and two book chapters. She has been a leader of four and participant of seven research projects, respectively. Currently, she holds a Postdoctoral research position at the Centre of Applied Economic Research, Faculty of Management and Economics, Tomas Bata University in Zlín.

Lorena G. Verde-Flota is Research Coordinator, Institute for Technology and Innovation in SMEs (IDIT*pyme*) at Universidad de Guadalajara (University of Guadalajara) and has a joint lecturing appointment at the InstitutoTecnológico de EstudiosSuperioresde Occidente (ITESO) Western Institute of Technology and Higher Education ITESO and Universidad de Guadalajara. She obtained her PhD in Social Scientific Studies from in Mexico, in the field of R&D management and technological innovation.

Takao Fujiwara is Professor of Business Administration, Department of Architecture and Civil Engineering, Graduate School of Engineering, Toyohashi University of Technology, Japan. He graduated in Economics from Aichi University and completed his Master's and PhD in the subject from Nagoya University. His research interests include biotech start-ups and real options.

Bernardo Jaén-Jiménez is Professor, University of Guadalajara, Economic and Administrative Sciences Campus. He holds a BA degree in Economics and a PhD in Social Sciences, both of them awarded by the University of Guadalajara. A member of the National Research System in Mexico, his fields of interest are international economy, macroeconomy, regional development and economy of innovation. His most recent publications are about the software industry and public policy in Mexico and innovation in small and medium enterprises (SMEs).

William H.A. Johnson is MBA Director at the Black School of Business and Associate Professor of Management, Pennsylvania State University, Erie. His research and teaching interests intersect innovation, strategy and international management, primarily revolving around the processes involved in the commercialization of new products and technologies. His research has been published in a number of leading innovation-based journals including, *Research Policy, IEEE Transactions on Engineering Management, Technovation, International Journal of Technology Management, Journal of International Management*, among others.

Konstantinos C. Kostopoulos is an Assistant Professor of Management at Norwich Business School (NBS), University of East Anglia, United Kingdom. Prior to joining NBS, he was an Assistant Professor of Management at EADA Business School, Barcelona, Spain. He holds a Master's degree in Business Administration and a PhD in Strategic Management, both fromthe Athens University of Economics and Business, Greece. He has performed research and advisory work in the fields of corporate strategy, human resource development, and R&D assessment and policy for large international organizations such as the European Commission, the SAS Institute, and the Athens International Airport. His work has appeared in leading scientific journals and conference proceedings including, among others, the *Journal of Management, Group & Organization Management*, and the *Journal of Business Research.*

Aline D. Masuda is Professor, EADA Business School, Spain. She holds PhD and Master's degrees in Industrial Organizational Psychology from the State University of New York-Albany and Missouri State University, respectively. Her research interests include motivation, job-satisfaction and well-being, work-family conflict and enrichment, global identity and culturally sensitive family-friendly policies. She has published articles in refereed academic journals such as *Journal of Vocational Behavior, Journal of Applied Psychology, An International Review*, *Small Group Research* and the *Journal of Business and Psychology.*

Giulia Monetta holds a PhD in Public Management, Salerno's University, Italy. Currently she is a temporary Researcher at University of Salerno's Department of Study and Business Research. Her research themes relate to strategic management and corporate governance of public and private firms. Her papers have been published in national and international journals.

Drahomira Pavelkova is working as a full-time professor at Tomas Bata University in Zlin, Faculty of Management and Economics, Czech Republic. She is also the dean of the faculty. Her areas of professional work and interest cover strategic management, clusters, corporate finance, performance management/ measurement, corporate valuation and benchmarking. She has presented at national and international conferences and published as author or co-author of monographs, chapters in books and articles in journals. She is a member of notable academic societies.

Lourdes Perez is Marketing Professor at Toulouse University, Toulouse Business School. She has been Adjunct Professor in Marketing at Pompeu Fabra University and Academic Dean at

EADA where she managed faculty and research. Her research work has been published in numerous articles and as case studies. Before becoming an academic, she was working in different multinational companies such as Hewlett Packard. She earned European PhD from the Technical University of Catalonia.

Vesa Routamaa is Professor of Management and Organization, University of Vaasa, Finland. He has over 200 publications that deal with issues relating to organizational structuring, managerial work, personality type, leadership, teams, creativity, innovations, change, cross-cultural issues, succession process and entrepreneurship. His company, Leadec Leadership Development Center Limited, carries out diverse research and training assignments in the fields of management, leadership, organizational development and entrepreneurship. He holds a DSc degree in Economics.

Berta E. Madrigal-Torres is Professor, University of Guadalajara and has specialized in leadership and entrepreneurship studies. He holds a Master's Degree in Management and a PhD in Administrative Sciences awarded by the National Polytechnic Institute. She has published more than seven books in her own subject areas as well as in the theme of human resource management and a number of papers in scientific journals. She is a member of Mexico's National Research System.

Ricardo Arechavala-Vargas is Founder and Director, Institute for Technology and Innovation in SMEs (IDIT*pyme*) at Universidad de Guadalajara; Professor and a member of the National Research System in Mexico. His consulting experience includes work on strategic technology management for public and private R&D labs and for private firms. As a researcher, his publications are in the fields of technology management, science and technology policy and the dynamics and development of technological capabilities in small and medium enterprises (SMEs).

Index